PATTERN RECOGNITION IN NANOSCIENCE, ENVIRONMENTAL ENGINEERING AND ARCHEOLOGY

PATTERN RECOGNITION IN NANOSCIENCE, ENVIRONMENTAL ENGINEERING AND ARCHEOLOGY

DIANE M. PEREZ
EDITOR

Nova Science Publishers, Inc.
New York

For permission to use material from this book please contact us:
Telephone 631-231-7269; Fax 631-231-8175
Web Site: http://www.novapublishers.com

LIBRARY OF CONGRESS CATALOGING-IN-PUBLICATION DATA

Pattern recognition in nanoscience, environmental engineering and archeology / Diane M. Perez, editor.
p. cm.
ISBN-13: 978-1-60021-715-9 (hardcover)
ISBN-10: 1-60021-715-X (hardcover)
1. Pattern formation (Physical sciences) 2. Nanoscience. 3. Environmental engineering. 4. Archaeology. I. Perez, Diane M.
Q172.5.C45P39 2006
500.201'185--dc22 2007019112

Published by Nova Science Publishers, Inc. ✦ New York

CONTENTS

PREFACE

This book presents new and important research on pattern recognition - the research area that studies the operation and design of systems that recognize patterns in data. It encloses subdisciplines like discriminant analysis, feature extraction, error estimation, cluster analysis (together sometimes called statistical pattern recognition), grammatical inference and parsing (sometimes called syntactical pattern recognition). Important application areas are image analysis, character recognition, speech analysis, man and machine diagnostics, person identification and industrial inspection.

An electronic nose (E-nose) is an analytical device used to detect and identify specific chemical components of complicated samples. Various E-nose systems have been used for a broad range of applications including medical diagnosis, pollutant and hazardous matter detection, environmental quality control, and food quality inspection. These instruments have the following advantages: (1) ability to continuously monitor, (2) ability to function in harsh environments, (3) accurate recognition specificity and enhanced sensitivity, and (4) autonomous functioning without human oversight.

Pattern recognition is a critical technique for an E-nose system, and it is important to establish reliable, robust and accurate detection and classification regimes. Considerable progress has been made in this research area, and this chapter presents a review of the pattern recognition algorithms that were used for novel E-nose sensors during the last five years. We also present ideas that readers may find useful for determining promising directions in sensor construction.

Chapter 1 divides pattern recognitions methods used for E-nose systems into three major categories: multivariate statistical analysis, neural networks, and other machine learning approaches. For each category, this chapter provides a relatively detailed framework or description for each method and illustrates its effects by various applications. Finally, while giving a retrospective summary, the authors discuss future perspectives of pattern recognition algorithms for constructing E-nose sensors.

The most important task in knowledge-based recognition procedure is the conversion of input process variables into features. The second is the translation of the classification logic expressed initially in natural linguistic form into a transparent recognition algorithm that yields from the structural variables membership to different classes. Pattern recognition is a potentially useful approach for interpreting the data generated by chemical processes. In contrast to expert system approaches to data interpretation where analysis proceeds deductively in a sequence of logical steps, pattern recognition involves inductive reasoning through generalization from a set of learned examples of process behavior. Pattern recognition may also complement the expert system approach by carrying out the low level assessment of process trends not easily captured in rules.

Wastewater processing by means of activated sludge is most widespread sewage treatment process. The activated sludge process (ASP) is with the further complication of operating in a widely changing environment where process conditions may be appreciably altered by immeasurable inputs.

Chapter 2 introduces the field of pattern recognition to the wastewater treatment plant in a much broader scope. This chapter analyses the theory involved in modeling the bioreactor (the major process equipments involved) followed by a comparative study of the IAWQ state-of-the-art model ASM1 with ASM3. It can be used as an introduction to the field in that it touches on the main branches of this discipline. No attempt will be made to introduce the reader to the broad field of artificial intelligence. Suffice to say that pattern recognition is a subset of artificial intelligence. This approach allows modeling and simulating an activated sludge process using its dynamic characteristics.

Traditional vision-based methods usually estimate motion or vibration from large sequence of unblurred images by employing optical flow algorithms on the image data. However, the image sequence tends to get blurred when the displacement of the object is large within the exposure time of the camera. In this chapter, the authors introduce a novel kind of approaches based on motion-blurred-images for measuring motion or vibration parameters. This group of approaches utilizes the motion blur information in the motion blurred images to measure motion or vibration parameters. Chapter 3 derives two classes of models to represent the motion blur information from the motion-blurred images and measure the motion or vibration parameters from the motion blur: the first one is based on image features in the spatial domain and the second one is based on that in the frequency domain.

In spatial domain, the authors employ geometric moments to extract the motion blur cue for measuring motion or vibration parameters. From the previous work, they derive a set of nonlinear equations of the motion parameters from the relationship between the geometric moments of translational motion blurred images and the displacement function of the motion. Thus, the authors calculate the parameters of the motion by solving these nonlinear equations.

In frequency domain, the spectrum of translational motion blurred image is multiplied by a Sinc function for translation or a zero-order Bessel function for high-frequency vibration, which has zeros at positions related to the length of the boxcar. When these zeros are found by image processing method, the parameters of image motions can be estimated.

For rotational motion, the authors represent the images in polar coordinate system. Obviously, estimation of the angular velocity of a monaxial rotation in polar coordinates is equivalent to estimating the velocity of 1D transitional motion in Cartesian coordinate system.

These techniques for measuring motion or vibration from motion blur can measure uniform motion, accelerated motion, vibration from only one motion blurred image and an unblurred image for reference or from only two successive frames of motion blurred images, which is almost impossible for high-speed video or stroboscopic photography. They employ examples with both synthetic and real motion blurred images to illustrate the performance of the proposed methods in estimating motion.

Recent development of nanoscience and nanotechnology has driven a tremendous research activity into tubular quasi-one-dimensional structures of nanometer size, such as carbon nanotubes, boron nitride nanotubes and many others, made of inorganic materials. More recently, in 2001 a novel kind of nanotubes was fabricated, namely, silicon carbide nanotubes (SiC–NTs), which are the subject of growing interest due to their potential applications in nanotechnology and their fundamental physical properties. From both perspectives, the study of optical response of SiC–NTs is of utmost importance. The authors report here a theoretical calculation of the linear optical coefficients (dielectric function, refractive index, reflectance and optical absorption coefficient) of single-walled SiC–NTs of different diameters, focusing on uniform ensembles of such tubes with a zigzag achiral structure. The calculation is carried out within a simple analytical model of the electronic structure of the SiC–NTs, based on an effective-mass approximation, and, at least to their knowledge, is the first for such nanotubes. The optical spectra we have calculated exhibit rich characteristic structures associated with interband electron transitions between pairs of van Hove singularities in the electronic density of states of the tubes. The collection of the results for the frequency-dependent linear optical coefficients, presented in Chapter 4, is a good test bed for the abovementioned model of the electronic structure of the SiC–NTs and, at the same time, may be useful for developing physical insights into the optical properties of such tubes.

As presented in Chapter 5, pattern recognition methods provide powerful tools for the analysis and interpretation of large environmental data sets generated within environmental monitoring programmes. Most of these data sets consist of trace elements and/or trace organic pollutants patterns. Only a few studies have been done on employing the pattern recognition methods to describe the correlation between locations and radioactive contamination. This multivariate approach has been used predominantly for the identification of radioactive isotopes, quantitative gamma-ray spectrometry analysis and for optimization of gamma-ray spectrometric measurements. Spatial data analysis based on radioactive contamination of diverse regions is complex for many reasons. These include the uncertainty of the source term, high spatial and temporal variability of pollution patterns, spatial and temporal non-stationary and the multivariate nature of the phenomenon with linearly and non-linearly correlated variables. The studies show that the geographic origin can be recognized with minimum effort if the relevant constituents are analyzed and the results are included in data analysis algorithms. Five common pattern-recognition techniques, artificial neural networks (ANN), principal component analysis (PCA), linear discriminant analysis (LDA), k-nearest neighbors (kNN) and soft independent modeling of class analogy (SIMCA) were employed to classify soil and bioindicator samples (mosses and lichens) according to their geographical origin, based on their content of radionuclides from different sources (members of the natural uranium and thorium decay chains, cesium isotopes originating from the Chernobyl power plant accident and cosmogenic beryllium), determined by gamma-ray spectrometry. The ability of the ANN to extract hidden features from the input signals was found to be particularly useful for this kind of monitoring, when the data sets with complex correlation structures had to be analyzed, or when data sets contained series of many highly inter correlated variables.

Pattern recognition is of primary importance to the field of archaeology, given that recognizing patterns is the key to both understanding problems with our current beliefs about the past and the generation of new and better ideas to resolve them. Chapter 6 discusses the use of multivariate statistics for data reduction and pattern recognition in large archaeological datasets. Most archaeological research is at least partially quantitative, but most of this research has traditionally focused on hypothesis testing and the calculation of probability statistics. While statistical hypothesis testing is an important aspect of scientific approaches to archaeological research, the field has largely overlooked the use of non-probability based statistical approaches to pattern recognition. The chapter begins with a brief discussion of the philosophy of science and the role of archaeological pattern recognition in the construction of accurate generalizations about the archaeological record based on material remains, the construction of hypotheses based on these generalizations, and the critical testing of hypotheses based on new and better understandings of archaeological patterning. This chapter then outlines the use of multivariate statistical approaches including factor analysis, principal components analysis, and multiple regression, which are powerful and under-utilized approaches to large archaeological datasets. The chapter illustrates the use of these techniques by discussing several important case study analyses of some major categories of materials and situations commonly encountered by archaeologists.

Chapter 7 deals with the problem of detecting in archaeological drawings a symbol whose shape is extremely variable. The authors describe both a segmentation and a recognition technique designed to solve this problem. The extraction of the symbols is based on encoding the drawing as a set of discrete curves, and on properly grouping these curves. Because of the great shape variability of the symbol, the authors propose a recognition technique based on a set of rules. These rules characterize the global morphology of the symbol while allowing large variations between its actual instances. One of the rules is the existence of a single symmetry axis. The authors describe a comparison metric between curves in order to search for symmetries. The other rules used for the recognition verify the presence of an arch at the core of the symbol. The authors present recognition results obtained on various pre-hispanic images. The rule-based approach proposed and implemented here appears suitable to detect polymorphic signs, a common feature of Mesoamerican symbols.

EXPERT COMMENTARY

In: Pattern Recognition in Nanoscience...
Editor: Diane M. Perez, pp. 3-5
ISBN: 1-60021-715-X

NON-LOCALITY IN PERCEPTION AND NANO-ROBOT NETWORKS

Fariel Shafee
Physics Department, Princeton University, Princeton, NJ 08540

The question of non-locality is embedded in pattern recognition. Human consciousness involves parallel processing of action potentials in different neurons in different areas of the brain. For example, even the recognition of a simple shape, such as a triangle or a circle, involves the simultaneous contribution of neurons at different positions in the visual cortex. Cognition of more complex nature would need correspondingly complex segmented localities of the cortex. Since it is unlikely that there is a master neuron which sequentially receives different kinds (frequencies) of signal from these widely separated segments and identifies the segments from that complex one-dimensional temporal pattern, we have to use at least two dimensional templates of excited neuron sets to resonate with specific perceptions, and the temporal sequence of these templates is the perception of events happening around us, or even our thoughts.

Non-locality is also a feature of quantum entanglement of states. Hence, entangled quantum networks in a sense are likely to represent most faithfully perceptions of living beings, or consciousness. Though this subject is the topic of much discussion [1,2,3,4,5], since consciousness cannot be precisely defined, it does point out the need for greater study of quantum neural networks. The large size of the biological neuron and the temperature of the brain question whether quantum coherence is a relevant factor [6] and usually we use classical models for signal processing in the brain for classical pattern recognition. However, the question of non-locality of perception remains intriguingly unanswered. As signals cannot travel faster than light, the idea of simultaneity is bounded by this limitation when combined with the spatial extent of the pattern template, if a non-entangled classical model is used. The action potential pulse has a finite width in time, though small, and that may have something to do with the classical simultaneity used by the brain in pattern recognition. A very thorough study of the quantities involved along with modeling equations that go beyond empirical Huxley-Hodgkin type [7] relations for action potential pulse propagation along an axon may be involved in such studies, which may lead to coupled equation systems producing a classical type of entanglement, and self organization, and that may in turn bring us nearer to the understanding of the biological concept of consciousness.

On the other hand if we invert the induction, biological models may suggest useful devices that are more efficient in pattern recognition. If such devices are small and can be treated semi-classically, then evolution suggests that it may be best if they are endowed with the property of being triggered at a certain level of excitation, so that noise from the environment and functionally unrelated neighbors may be discarded. Such semiclassical nets [8] may be composed of nanomachines working as the neuron nodes. They may be dispersed in the matrix of the system components according to a geometric distribution that may accentuate the region of interest. In general this distribution may also be temporal too, to investigate time series patterns, or even in momentum phase space, to get position-independent pattern information as Fourier transform of the configuration space.

At a higher level of purity and quantum coherence, where entanglement may be significantly high, with only a small amount of impurity in the system density matrices, we may hope to use these nano-robots to signal coherent correlated information modified by local characteristics which may be back-projected to detect the presence of a pattern. Since this will be a finite set of robots, we cannot get a full image as in a hologram, which is exactly the purpose of pattern recognition, ignoring irrelevant differences, but categorizing the sample on the basis of a subset of parameters. A systematic procedure for the design of such a social network [9,10] of nano-robots may be a problem worth looking into.

These nano-robots, in turn lead us back to the possibility of using such devices in a biological context of great importance. Let us assume that such devices at the molecular or nano scale are provided with entangled quantum markers, more general than simple qubit spins, with many (N) possible states, and that initially the quantum state of the system of n markers is

$$|\varphi> = (1/\sqrt{N}) \sum_{i=1}^{N} \prod_{j=1}^{n} |i,j>$$

Now, if one of the itinerant nano-robots come sufficiently near a target specimen in state |I> and forms a bound state with it, that would trigger the collapse of that nano-robot to the i-th state, the, because of entanglement, all other nano-robots in the aforementioned domain of influence will collapse to the i-th state. The system can be, in theory, such that nano-robots in the same state attract, similar to parallel-spin objects. In this case there may be a convergence of the nano-robots to the (i-I) target-nanorobot bound system engulfing and disabling it completely. This may have interesting implications in developing more effective immunological reactions at a quantum level.

References

[1] Penrose, R. *The Emperor's New Mind.* (Oxford U.P., 1989)

[2] Penrose, R. *Shadows of the Mind.* Oxford U.P., 1994)

[3] Honderich, T. 1984, The time of a conscious sensory experience and mind-brain theories, *J. Theoretical Biology* **110**, 115-29 (1984)

[4] Popper, K R. and Eccles, J.C., *The Self and Its Brain* (Berlin: Springer-Verlag, 1977)

[5] Schrödinger, E., *Mind and Matter* (Cambridge U.P., 1958) p. 37

[6] Tegmark, M., The Importance of Quantum Decoherence in Brain Processes, *Phys. Rev. E* **61** (2000) 4194-4206

[7] Hodgkin, A.L. and A.F. Huxley, A.F., *J. Physiol.* **117**, 500-540 (1952).

[8] Shafee, F., *Semiclassical Neural Network*, quant-ph/0202015 (2002)

[9] Shafee, F., A *Spin Glass Model of Human Logic Systems*, physics/0509065 (2005)

[10] Shafee, F., *Spin-glass-like Dynamics of Social Networks*, physics/0506161 (2005)

RESEARCH AND REVIEW STUDIES

In: Pattern Recognition in Nanoscience...
Editor: Diane M. Perez, pp. 9-41
ISBN: 1-60021-715-X

Chapter 1

Pattern Recognition Algorithms to Analyze Data Outputs for Novel E-Nose Sensors

Weixiang Zhao and Cristina E. Davis
Department of Mechanical and Aeronautical Engineering,
University of California, Davis, Davis CA 95616

Abstract

An electronic nose (E-nose) is an analytical device used to detect and identify specific chemical components of complicated samples. Various E-nose systems have been used for a broad range of applications including medical diagnosis, pollutant and hazardous matter detection, environmental quality control, and food quality inspection. These instruments have the following advantages: (1) ability to continuously monitor, (2) ability to function in harsh environments, (3) accurate recognition specificity and enhanced sensitivity, and (4) autonomous functioning without human oversight.

Pattern recognition is a critical technique for an E-nose system, and it is important to establish reliable, robust and accurate detection and classification regimes. Considerable progress has been made in this research area, and this chapter presents a review of the pattern recognition algorithms that were used for novel E-nose sensors during the last five years. We also present ideas that readers may find useful for determining promising directions in sensor construction.

This chapter divides pattern recognitions methods used for E-nose systems into three major categories: multivariate statistical analysis, neural networks, and other machine learning approaches. For each category, this chapter provides a relatively detailed framework or description for each method and illustrates its effects by various applications. Finally, while giving a retrospective summary, we discuss future perspectives of pattern recognition algorithms for constructing E-nose sensors.

1. Introduction

Surprisingly, one of the best sensors in existence is the canine nose. It is both specific and sensitive to many compounds of interest, both chemical and biological. Currently, experienced and well-trained police dogs are able to detect toxic substances, narcotics and even human scents. However, sometimes background smells in environments can obscure certain scents, making detection for the dogs difficult. Also, the dogs require long rest breaks to prevent failure of detection, and they need very experienced trainers and handlers. For these and many other reasons, scientists and engineers have sought to develop an electronic olfactory organ, called electronic nose or E-nose, (illustrated in Figure 1). These sensors were specifically invented to detect and identify specific chemical components in challenging environments, and there are many different sensing modalities that comprise this class of instruments. During the past 20 years, E-nose systems have been significantly improved and widely applied to various applications because of their following advantages: (1) ability to continuously monitor, (2) ability to function in a harsh environment, (3) accurate recognition specificity and enhanced sensitivity, and (4) autonomous functioning without human oversight.

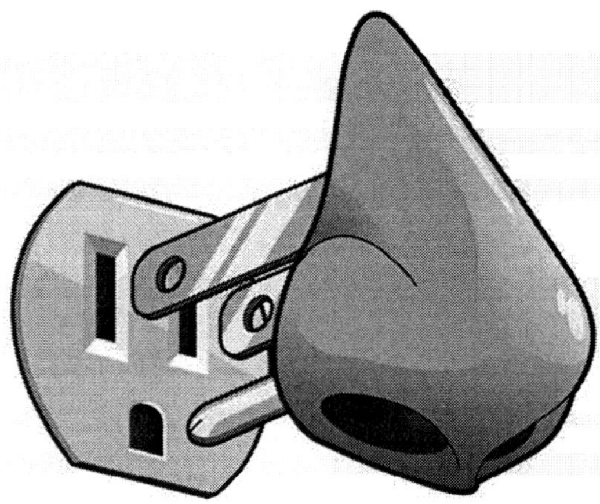

Figure 1. An electronic device with olfactory function (Minkel, 2005). (Reprinted with permission from *The Scientist*; Illustration: Brad Fitzpatrick)

Among the many application areas of interest, E-noses have been utilized for space exploration, cosmetics industry, safety protection, and material engineering. The successes of E-noses have been obtained in monitoring air contaminants in spacecraft and space shuttle (Qian *et al.*, 2006; Young *et al.*, 2003), discriminating perfume compound in fragrance (Branca *et al.*, 2003), detecting fires for safety protection (Scorsone *et al.*, 2006), and monitoring the production of organic semiconductors for thin-film-transistors (Torsi *et al.*, 2000).

In the medical science field, E-nose systems have been employed to detect lung cancer based on virtual surface acoustic wave (SAW) gas sensors (Chen *et al.*, 2005), to classify traditional herb medicines (Sim *et al.*, 2003), to detect ethanol release from sol-gel bioencapsulation (Lovino *et al.*, 2005), to diagnose illness (Gardner *et al.*, 2000b), and to discriminate and classify clinical anaerobic isolates (Pavlou *et al.*, 2002, Dutta *et al.*, 2006).

In the field of environmental quality protection, E-nose systems have been successfully applied to detect and classify chemicals for spacecraft environmental monitoring (Qian *et al.*, 2006), to identify bacterial infections in hospital environments (Dutta *et al.*, 2005), to determine the relationship between sewage odor and biochemical oxygen demand (Onkal-

Engin *et al.*, 2005a), to classify sewage odor (Onkal-Engin *et al.*, 2005b), to monitor the exhaust air of a compost pile (Romain *et al.*, 2005), and to assign malodors to environmental sources (Nicolas *et al.*, 2000).

Another major application of E-noses is food quality control. Some of the typical applications included detection and differentiation between *Streptomyces* in potable water (Bastos and Magan, 2006), food vapor recognition (Yang *et al.*, 2005), cigarette brand identification (Luo *et al.*, 2004), discrimination among milk and cultured dairy products (Collier *et al.*, 2003), study of shelf life and cultivar effect on tomato aroma profile (Berna *et al.*, 2004), quantification of spice mixture compositions (Zhang *et al.*, 2005), detection of rancidity in potato crips (Vinaixa *et al.*, 2005a), and white wine classification (Cozzolino *et al.*, 2005).

E-noses are being given increased attentions because of their distinctive advantages. Figure 2 shows the major modules of natural (biological) and electrical odor identification processes. It is clear that a good E-nose system relies on two major parts: complete collection of sensor signals and correct techniques for data analysis. For E-nose based classifiers, data analysis specifically means pattern recognition techniques. Technically speaking, the goal of pattern recognition is to classify patterns from *a priori* information. With the improvement in computer technology, pattern recognition has become a powerful machine learning tool to discover how we can categorize complex data sets. Therefore in this chapter, we will review as completely as possible the pattern recognition algorithms that were used for E-nose systems during the last five years, provide a retrospective summary, and discuss future perspectives of pattern recognition algorithms for constructing E-nose systems. We hope this chapter can present ideas that readers may find useful for determining promising directions for their E-nose system construction.

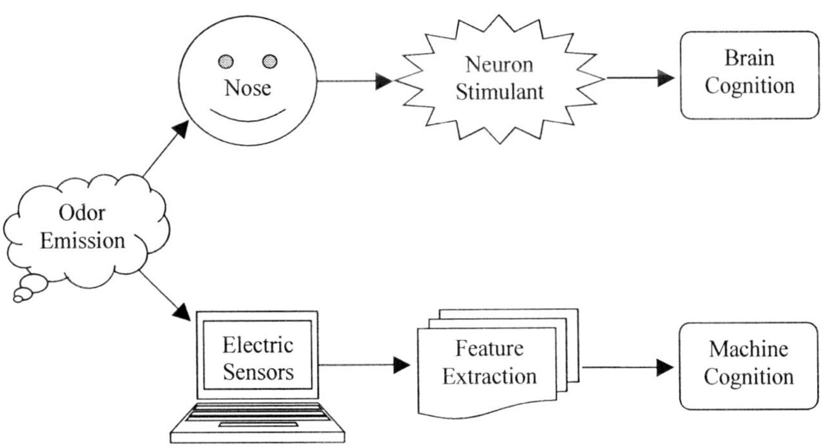

Figure 2. Natural and electrical odor identification processes.

2. Pattern Recognition Algorithms for E-Noses

In general, pattern recognition is organized in the context of supervised learning techniques. Figure 3 shows a basic framework of supervised learning. The basic concept of pattern recognition is to find a map (relationship) from sample features to sample class index and then this map can be used to predict the class of a new unknown sample.

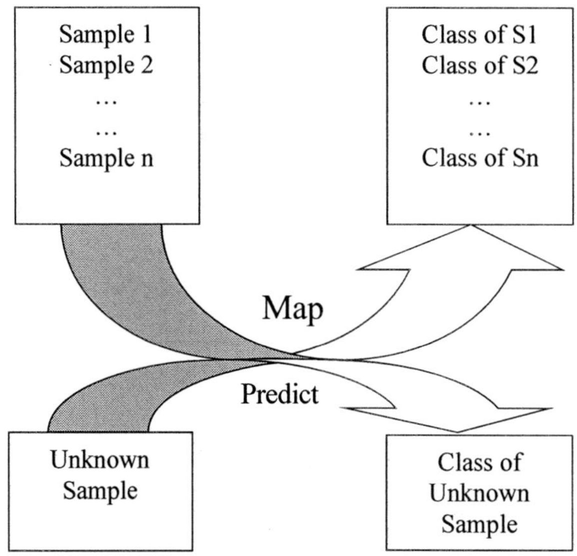

Figure 3. Framework of supervised learning.

A typical pattern recognition system usually is composed of four modules:

(1) **Data collecting**. In this step, sufficient samples are collected and each sample is composed of a number of physical variables that characterize the observed system.

(2) **Data pre-processing**. In this step, we may want to filter noise, normalize data, or physically align the sensor output data. These actions may enhance signals and eliminate possible obscuring effects among variables.

(3) **Feature extracting**. For some systems, samples are composed of an extremely large number of variables. A complex chemical industrial operating unit could be controlled by hundreds of physical variables. A high resolution mass spectrum vector could be composed of thousands of m/z mass-to-charge signals. A comprehensive environmental monitoring system may need to collect information on hundreds of chemicals. It is therefore necessary to use feature extracting methods to transfer these data from their original high dimensional space to lower dimensional feature space. Auto-regression modeling, cepstrum analysis, and principal component analysis are some common feature extracting methods.

(4) **Classification**. Various classification strategies (supervised learning methods) are applied to the feature space to create a map (relationship) from sample vectors to sample class indexes. This step is the core part of the entire pattern recognition system and is also the major component of this chapter.

In this section, various pattern recognition algorithms for supervised learning are categorized into three groups: multivariate statistical analysis methods, neural network methods, and other machine learning methods. A description of each method will be given along with examples of E-nose application case studies. Some unsupervised learning algorithms will also be briefly introduced here, as they are often used for data pre-processing or feature extracting for supervised learning.

2.1. Multivariate Statistical Analysis Methods

Multivariate analysis is the most canonical tool for modeling a complex system. It has the longest development history and has been extensively employed for pattern recognition.

2.1.1. Principal Component Analysis (PCA)

A. Principle

PCA is a linear transformation that transforms original data to a new coordinate system in which the greatest variance of the data lie on the first coordinate (called the first principal component), the second greatest variance on the second coordinate, and so forth. In other words, PCA transforms correlated variables into a fewer independent variables called principal components. Therefore, with the ability to reduce system dimensions and produce new underlying latent variables, PCA is widely used as a feature extracting method for pattern recognition in E-nose systems (Vinaixa *et al.*, 2005b; Marti *et al.*, 2004; Strathmann *et al.*, 2005; Hamilton *et al.*, 2005; Rocha *et al.*, 2006; Presicce *et al.*, 2006; Carmona *et al.*, 2006, Prasad *et al.*, 2006).

The mathematical description of the PCA transformation can be briefly shown as below. Suppose the original system can be represented by a *p*-dimensional vector, $\mathbf{x} = (x_1, x_2, ... x_p)$. The new coordinate system, f_1 , f_2, ..., f_m, $m \le p$, can be generated by eq. (1).

$$\begin{cases} f_1 = a_{11}x_1 + a_{12}x_2 + ... + a_{1p}x_p \\ f_2 = a_{21}x_1 + a_{22}x_2 + ... + a_{2p}x_p \\ \vdots \\ f_m = a_{m1}x_1 + a_{m2}x_2 + ... + a_{mp}x_p \end{cases} \tag{1}$$

where, $\text{cov}(f_i, f_j) = 0$, $i \neq j$. This indicates any two new coordinates are independent.

Let the mean vector and the covariance matrix of the whole pattern set $\{\mathbf{x}\}$ be $\boldsymbol{\mu}$ and $\boldsymbol{\Sigma}$, respectively. According to matrix theory, $\boldsymbol{\Sigma}$ can be decomposed as:

$$\boldsymbol{\Sigma} = \mathbf{U}\begin{bmatrix} \lambda_1 & & & \\ & \lambda_2 & & \\ & & \ddots & \\ & & & \lambda_p \end{bmatrix}\mathbf{U}^T = \sum_{i=1}^{p} \lambda_i \mathbf{u}_i \mathbf{u}_i^T \tag{2}$$

where, λ_i is the ith eigen value of $\mathbf{\Sigma}$ ($\lambda_1 \geq \lambda_2 \geq ... \geq \lambda p \geq 0$) and $\mathbf{u}_i$ is the ith eigen vector. It can be proved that $f_1 = \mathbf{u}_1^T \mathbf{x}$ is the first principal component of the whole pattern set $\{\mathbf{x}\}$, $f_2 = \mathbf{u}_2^T \mathbf{x}$ is the second principal component, and so on. This way, the original pattern set is transformed to a new coordinate system in which any two coordinates are independent. In most cases, the first few principal components are sufficient to reflect the whole system and used for further analysis. A conventional way to determine the proper number of principal components is based on the accumulated variance contribution ratio, which can be expressed as:

$$\frac{\sum_{i=1}^{m} \lambda_i}{\sum_{i=1}^{p} \lambda_i} > \text{Threshold (e.g. } 0.95) \quad (3)$$

Figure 4 illustrates a typical principal component transformation on a two dimensional space. The length of the new coordinate represents the variance on this direction. As an effective dimension reduction method, PCA is the most widely used data pre-processing method for E-nose pattern recognition.

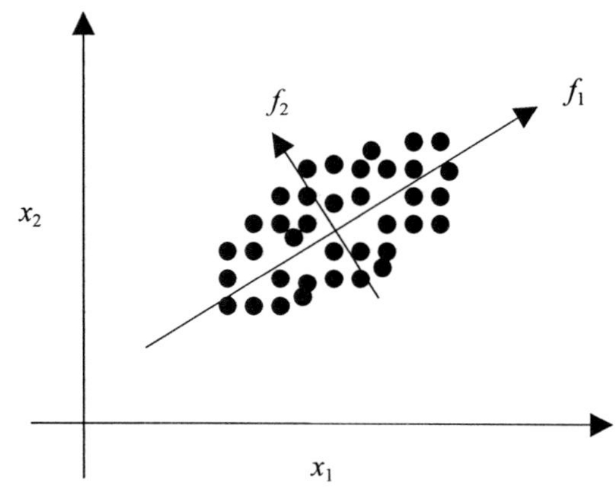

Figure 4. Principal component transformation on a two dimensional space.

B. Case Studies

In general, an E-nose system is composed of a large number of E-sensors and the features represented by these sensors are very likely to be correlated with each other. Therefore, PCA often becomes the first choice for data pre-processing in E-nose systems. More than half of the E-nose studies during the last five years reported the usage of PCA for dimension reduction, variable noise removal, and visible classification.

- In an E-nose based wine classification study (Marti *et al.*, 2004), PCA was applied to classify wines into different origins (Priorat and Terra Alta), different varieties (Cabernet sauvignon, Tempraonillo and Merlot), and different aging statuses (Young wines, Crianza wines, Reserva wines, and Gran Reserva wines) based on the gas chromatography mass spectrometry (GC/MS) data of the volatile chemicals in the headspace of the wine samples.
- PCA showed a good visible classification result in an E-nose based combustion detection study (Scorsone *et al.*, 2006). In this study, four types of fires (smoldering cotton, smoldering wood, smoldering paper, and flaming plastic) and cigarette smoking were employed to test the feasibility of the detection strategy. The original inputs of the classification system were the GC/MS data of the collected fire emissions or smoke.
- PCA was successfully applied to the study of early detection of fungal growth in bakery products by a mass spectrometry based E-nose (Vinaixa, *et al.*, 2004). Seven fungal species (*Aspergillus flavus*, *Aspergillus niger*, *Eurotium amstelodami*, *Eurotium herbariorum*, *Eurotium rubrum*, *Eurotium repens*, and *Penicillium corylophillum*) were isolated from bakery products and used for the study. The first two principal components accounted for 88% variance of the data. PCA not only provided a visible separation of *E. amstelodami* and other *Eurotium* species but also produced a dimension-reduced data matrix for the subsequent fuzzy ARTMAP based fungal genus detection.
- PCA also showed its satisfactory applications in the study of shelf life and cultivar effect on tomato aroma profile (Berna *et al.*, 2004) and the analysis of saffron volatile fraction (Carmona *et al.*, 2006).

2.1.2. *k*-nearest Neighbors (*k*-NN)

A. Principle

k-NN is a relatively straightforward pattern recognition method and has been effectively used for E-nose sensors (Dodd *et al*, 2004; Kuske *et al.*, 2005). Assume there are N samples of C classes, in order to classify an unknown sample x, *k*-NN needs to find k nearest neighbors of this new sample among N samples in terms of distance or similarity. Let the number of the nearest neighbor samples from classes 1, 2, ..., and C be k_1, k_2,..., and k_C, respectively $(k_1 + k_2 + \cdots + k_C = k)$. Then, the decision rule is:

$$\mathbf{x} = \arg\max_j : k_j \quad (j = 1,2,...,C) \tag{4}$$

The number of the nearest neighbors from each known class determines the class index of an unknown pattern. The key parameter for this method is k, which depends on data. A larger value of k could reduce the influence of outlier or noise but could generate clouded class boundaries.

B. Case Studies

- *k*-NN was used for a food decay study examining Tilapia (seafood) storage time classification (Dodd *et al.*, 2004). In this study, 16 tin metal oxide sensors with various chemical sensitivities were employed to collect signals for the E-nose classifier. An 18-hour accelerated decay period was divided into three 6-hour time regions. The goal of this study was to detect the decay status (one of three time regions) of decaying Tilapia. *k*-NN showed better classification results than the conventional least square regression method, with the best classification rate being 97.8% for *k*-NN.

2.1.3. Linear Discrimination Analysis (LDA)

A. Principle

LDA is a linear classifier based on an assumption that patterns of different classes are linear separable. It is extensively used in E-nose systems (Buratti, *et al.*, 2004; Yu and Wang, 2007; Lozano *et al.*, 2006; Kuske *et al.*, 2005). Let us start to describe this method using a two-class separation problem. As shown in Figure 5, the goal of LDA is find a proper line to separate two groups of patterns.

Let a pattern vector be $\mathbf{x} = (x_1, x_2, ...x_p)$. A general equation for LDA is:

$$f(\mathbf{x}) = \mathbf{w}^T\mathbf{x} + w_0 \tag{5}$$

where, w is a *p*-dimensional weight vector and w_0 is a constant or a threshold value. The decision rule for a two-class problem can be expressed as:

$$\mathbf{x}\begin{cases} \in Class\,1 & \text{if } f(\mathbf{x}) > 0 \\ \in Class\,2 & \text{if } f(\mathbf{x}) < 0 \end{cases} \tag{6}$$

The key for designing a LDA classifier is to find the best separation line (i.e., the best w and w_0) from the training samples. Fisher discrimination (Fisher, 1936) is a canonical optimization strategy for finding the best parameters w and w_0. The key concept of Fisher discrimination is to maximize the ratio of the between-to-within variability of the samples from two classes.

Let the mean vector of each class be $\mathbf{q}_i = (1/N_i)\sum_{k\in i}\mathbf{x}_k \; (i = 1,2)$. The between- and within-covariance matrices (S_B and S_W) of the whole sample set can be expresses as:

$$S_B = (\mathbf{q}_1 - \mathbf{q}_2)(\mathbf{q}_1 - \mathbf{q}_2)^T \tag{7}$$

$$S_W = \sum_{i=1}^{2}\sum_{k\in i}(\mathbf{x}_k - \mathbf{q}_i)(\mathbf{x}_k - \mathbf{q}_i)^T \tag{8}$$

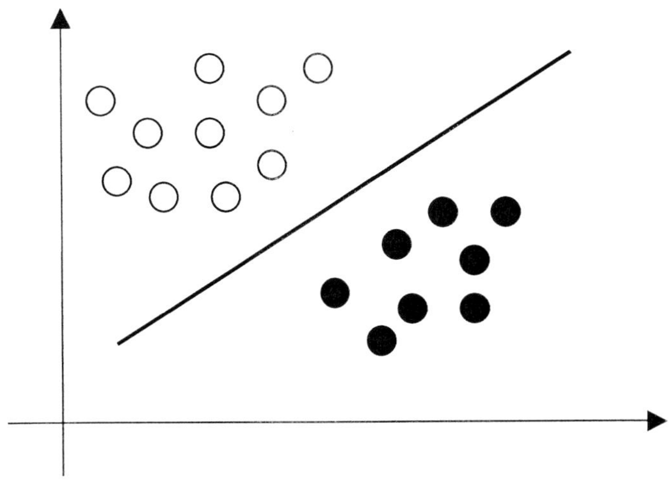

Figure 5. Illustration of a two-class linear separation.

The Fisher rule can be derived into the following optimization equation. Please refer to (Fisher, 1936) for the detailed deduction.

$$\max J(\mathbf{w}) = \frac{\mathbf{w}^T S_B \mathbf{w}}{\mathbf{w}^T S_W \mathbf{w}} \tag{9}$$

The solution of this optimization problem is $\mathbf{w}^{opt} = S_W^{-1}(\mathbf{q}_1 - \mathbf{q}_2)$ and the corresponding w_0^{opt} can be easily calculated with $\mathbf{w}^{opt}$. Theoretically speaking, $\mathbf{w}^{opt}$ and w_0^{opt} yield the best separation boundary between two groups. Another attractive feature of discrimination analysis is to permit one to observe the spatial distribution of different groups based on two or three discriminant scores.

In addition to the Fisher discrimination rule, the perceptron rule (which is recognized as the base of the development for artificial neural networks) is also an effective strategy to determine the best linear separation parameters.

The above two-class decision rules can be extended to a multiple-class separation problem. For a C-class separation problem, the upper limit number of decision planes is $C(C-1)/2$, but in many cases one may need a much smaller number of decision planes. Stepwise linear discrimination functions are a powerful tool for multiple-class separation problems.

B. Case Studies

LDA is also a very popular pattern recognition method for E-nose systems, because of its simplicity and efficiency.

- LDA was employed for an E-nose and an E-tongue to classify Italian Barbera wines (Buratti, *et al*., 2004). The entire classification system contains 28 variables: 10 metal oxide semiconductor chemical sensors to form an E-nose array, 5 sensors to form an E-tongue array, 2 variables for spectrophotometric evaluation of color, and 11 chemical parameters. This entire system was tested on 53 wine samples of four types: 23 *Barbera Oltrepo*, 13 *Barbera Piemonte*, 12 *Barbera Astp*, and 5 *Barbera Alba.* Classifiers were built and tested to evaluate the classification ability of LDA on these samples. When only E-nose and E-tongue data were employed, LDA showed the best classification rate, 100% correct assignation and 98.1% correct prediction.
- Another example of the application of LDA was also in the wine industry (Cozzolino et al., 2005). In this study, the components in the sample headspace were directly passed to the mass spectrometer (MS) detector. The MS data were used as inputs for LDA, and the goal was to classify two types of Australian white wines, Riesling and unwooded Chardonnay. PCA was first applied to reduce the sample dimension. The first two principal components accounted for 79% of the sample variance. Then LDA employed the first two principal components and generated a 73% classification rate for unwooded Chardonnay and 82% for Riesling wines. Because of the information loss caused by employing only the first two principal components, the classification accuracy of LDA was poorer than discriminant partial least squares regression (DPLS). The detailed results of DPLS will be discussed later in this chapter.

2.1.4. Maximum Probability Rule

A. Principle

An important feature of a good classifier is the ability to minimize the "expected risk" caused by misclassification. Maximal probability rule based on Baye's theorem (Dutta and Dutta, 2006) is such a classifier.

Assume there are N classes, Class *1,2,.., N*. The posterior probability of class h, given sample x, can be represented as:

$$P(h/\mathbf{x}) = \frac{\lambda_h P(\mathbf{x}/h)}{\sum_{j=1}^{N} \lambda_j P(\mathbf{x}/j)} \tag{10}$$

where, λ_j is the priori probability of class j and $P(\mathbf{x}/j)$ is the conditional probability of x given class j. The maximum probability decision rule is:

$$\mathbf{x} = \arg\max_h [P(h/\mathbf{x})] \qquad (h = 1,2,...,N) \tag{11}$$

The conditional probability $P(\mathbf{x}/h)$ in eq. (10) can be replaced by a continuous probability density function $f(\mathbf{x}/h)$ in most cases. The priori probability λ_h can be estimated as the ratio of the number of the samples in Class h to the number of all the samples. Then the problem is how to estimate the probability density function $f(\mathbf{x}/h)$. Generally, there are two methods for this estimation, namely the parametric approach and non-parametric approach (Dutta and Dutta, 2006).

In the parametric approach, the conditional probability density function $f(\mathbf{x}/h)$ can be estimated by:

$$f(\mathbf{x}/h) = \frac{1}{\sqrt{(2\pi^{p})}\sqrt{|\boldsymbol{\Sigma}_h|}} \exp[-\frac{1}{2}(\mathbf{x}-\boldsymbol{\mu}_h)^T \boldsymbol{\Sigma}_h^{-1}(\mathbf{x}-\boldsymbol{\mu}_h)] \quad (12)$$

where $\boldsymbol{\Sigma}_h$ is the covariance matrix of class *h*, $\boldsymbol{\mu}_h$ is the mean vector of class *h*, and *p* is the pattern dimension. Thus, eq. (10) becomes

$$P(h/\mathbf{x}) = \frac{\lambda_h \frac{1}{\sqrt{(2\pi^{p})}\sqrt{|\boldsymbol{\Sigma}_h|}} \exp[-\frac{1}{2}(\mathbf{x}-\boldsymbol{\mu}_h)^T \boldsymbol{\Sigma}_h^{-1}(\mathbf{x}-\boldsymbol{\mu}_h)]}{\sum_{j=1}^{N} \lambda_j \frac{1}{\sqrt{(2\pi^{p})}\sqrt{|\boldsymbol{\Sigma}_j|}} \exp[-\frac{1}{2}(\mathbf{x}-\boldsymbol{\mu}_j)^T \boldsymbol{\Sigma}_j^{-1}(\mathbf{x}-\boldsymbol{\mu}_j)]} \quad (13)$$

For pattern recognition, only the numerator is useful for decision, as the denominator is the same for every class. The probability density function in eq. (12) is quadratic in terms of x, so this type of probability decision rule is also called quadratic discrimination (QD) technique (Dutta and Dutta, 2006).

In the non-parametric approach, kernel functions are used to estimate the conditional probability density function $f(\mathbf{x}/h)$. One type of kernel function (Dutta and Dutta, 2006) is:

$$f(\mathbf{x}/h) = \frac{1}{N_h}\sum_{i=1}^{N_h}\prod_{j=1}^{P}\frac{1}{\sigma_h}\exp\left[-\frac{1}{2}\left(\frac{\mathbf{x}_j-\mathbf{x}_{i,j}}{\sigma_h}\right)\right] \quad (14)$$

where, N_h is the number of the training samples of class h and σ_h is the smoothing parameter for class *h*. Kernel functions are also widely used for neural network construction, which will be discussed later in detail.

B. Case Studies

- The maximum probability rule was successfully applied for an E-nose to classify bacterial infections in a hospital environment (Dutta and Dutta, 2006). In this study, an E-nose comprising an array of 32 polymer carbon black composite sensors was used to collect the signals for classifying three types of bacterial infections: methicillin-resistant *Staphylococcus aureus* (MRSA), methicillin-susceptible *S. aureus* (MSSA), and coagulase-negative *Staphylococci* (C-NS). The sample set was composed of 600 infection samples from 600 patients in the hospital, each type of infection contributing 200 samples. This study compared the recognition effects for three probability rule based methods (QD classifier, kernel estimator classifier, and adaptive kernel estimator classifier). Two non-parametric classifier (i.e., the kernel estimator classifiers) showed better class separation than the parametric classifier (QD classifier). The adaptive kernel method showed the best recognition accuracy of 99.8%, which could indicate the better ability of the adaptive technique to estimate tail probability of the data distribution.

2.1.5. Partial Least Square Regression (PLSR)

A. Principle

If indexes of different classes are given different values (for example, "1" for class 1 and "0" for class 2), pattern recognition becomes a functional approximation (function map) problem. The inputs for this functional approximation are sample features and the corresponding output is the class index of this sample. Thus regression models are a simple but effective tool for this functional approximation.

Let us start describing this method with a two-class linear separable problem. Assume the sample set and the corresponding class indexes are $\{\mathbf{x}_i \mid i = 1,2...,N\}$ and $\{y_i = 1\, or\; 0 \mid i = 1,2...,N\}$, respectively. The goal of pattern recognition is to find a linear function for the map $y_i = f_{linear}(\mathbf{x}_i)$. Clearly, conventional least square regression (LSR) could be a choice for building this function map. In many case, however, conventional LSR may encounter a severe problem due to colinearity among variables and possible noise contained in samples. An extreme colinearity example is that one variable is a linear combination of some of other variables. A severe colinearity could result in the ill-condition of the regression matrix and further lead to the total failure of regression modeling.

Partial least square regression (PLSR) is a modified LSR. Like PCA, PLSR also extracts uncorrelated latent variables from the original sample features. The difference between PLSR and the principal component regression (PCR) is the extraction of PLS component employs the information of both independent variables (features) and dependent variables. The PLSR process is briefly described below.

Let X and Y denote independent and dependent data matrices. A general linear model is $\mathbf{Y} = \mathbf{XB} + \mathbf{E}$. PLSR first generates a transforming matrix W to transform X to T ($\mathbf{T} = \mathbf{XW}$, called score matrix or PLS components). Then conventional LSR is used to estimate the regression matrix M of T against Y, $\mathbf{Y} = \mathbf{TM} + \mathbf{E}$. Now the original regression model turns to be $\mathbf{Y} = \mathbf{XWM} + \mathbf{E}$. Clearly, $\mathbf{B} = \mathbf{WM}$. Thus, the model can be used for prediction. The key is to get the transforming matrix W to extract the correlated information between X and Y as much as possible. One standard PLS algorithm is nonlinear iterative partial least squares (NIPALS). The detail of this algorithm can be referred to (Hoskuldsson, 1988). For a C-class problem, the output of each sample can be denoted by a C dimensional vector (0,0,...1,...,0), which means this sample belongs to the class marked by "1".

With the ability to eliminate the colinearity among feature variables and to simultaneously model multiple dependent variables, PLSR has become a very effective linear function approximation tool and has been widely applied to E-nose system (Vestergaard *et al.*, 2006; Trihaas *et al.*, 2005a; Trihaas *et al.*, 2005b; Aishima, 2004).

B. Case Studies

- In the Australian white wines classification study previously described, in addition to the LDA (section 2.1.3), researchers also applied discriminant PLS (DPLS, an extension of PLS) for this E-nose classification problem and had high classification accuracies (>90%) for both Riesling and unwooded Chardonnay wines. This was better than the LDA method based on the first two principal components. One major reason for this improvement was that the comprehensive (latent) factors extracted by PLS reflected the variance information of both dependent and independent variables. Therefore in general, it is expected that PLS should have a slightly higher modeling accuracy than principal component based techniques.
- In addition to the successful applications to E-nose based pattern recognition, PLS was applied to other E-nose functions. PLS was used to correlate E-nose sensory attributes with GC/MS profiles (Aishima, 2004), to predict the crisp rancidity (Vinaixa *et al.*, 2005b), and to estimate the concentrations of the major chemicals causing boar taint (odor) of entire male pigs (Vestergaard *et al.*, 2006). The prediction of crisp rancidity will be discussed in detailed later along with the application of genetic algorithms.

2.2. Artificial Neural Networks (ANN)

The methods introduced earlier in this chapter are mainly used for linear classification. Although some of their modifications, such as stepwise linear discrimination function and polynominal regression, can be applied for non-linear classification, the most successful nonlinear classification tool is an artificial neural network (ANN). ANN is a good mimic of human brain cognition systems. It is able to classify samples with complex and nonlinear space distribution by a self-learning process in which knowledge and information are stored in the connecting weights between neurons. The following sections will describe the basic principles of various ANN methods used for E-nose systems.

2.2.1. Back-Propagation Neural Network (BPNN)

A. Principle

BPNN is one of the most widely used neural networks. It has obtained extensive and successful applications in function approximation, given that in most cases pattern recognition can be considered as a type of function approximation. A typical structure of BPNN is shown in Figure 6.

Assume a three-layer NN is composed of p input nodes, n hidden nodes, and q output nodes. Let a learning sample k be $\mathbf{a}_k = (a_1, a_2, \cdots, a_p)$ and its corresponding stipulated output be $\mathbf{y}_\mathbf{k} = (y_1, y_2, \cdots y_q)$. Also let the weights between input and hidden layers be

$\mathbf{w}_{ij}, i=1,2,\cdots,p, j=1,2,\cdots n$, and the weights between hidden and output layers be $\mathbf{v}_{jt}, j=1,2,\cdots,n, t=1,2,\cdots,q$. The most important element to construct a single neuron is the activation function that is used to produce the output response of each neuron. The Sigmoid function (S-function) is a standard activation function for each hidden and output neuron, but in some cases one nonlinear layer (the layer composed of nonlinear activation neurons) is accurate enough for approximation.

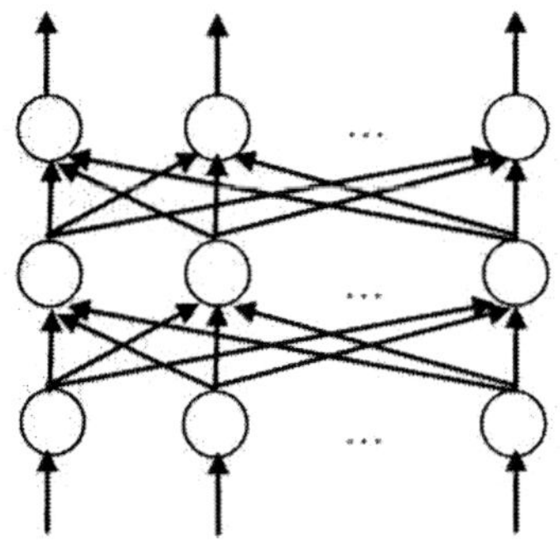

Figure 6. Structure of a three-layer BPNN.

Assume the total error between the model and stipulated outputs for learning sample k is

$$E_k = \sum_{t=1}^{q} (\mathbf{y}_{k,t} - \mathbf{c}_{k,t})^2 / 2 \tag{15}$$

where $\mathbf{c}_{k,t}$ is the model output, $t=1,2,\cdots,q$. The adjustments of the weight $\mathbf{v}_{jt}$ and $\mathbf{w}_{ij}$ caused by the error of sample k can be expressed as

$$\Delta \mathbf{v}_{jt} = -\alpha \left[\frac{\partial E_k}{\partial \mathbf{v}_{jt}} \right] \quad j=1,2,\cdots n, \ t=1,2,\cdots,q \tag{16}$$

$$\Delta \mathbf{w}_{ij} = -\beta \left[\frac{\partial E_k}{\mathbf{w}_{ij}} \right] \quad i=1,2,\cdots,p, \ j=1,2,\cdots n \tag{17}$$

where, α and β are the learning rates.

Similarly, the threshold value of each neuron can be adjusted by propagating the error backward from the output layer. This adjusting process is performed for each learning sample, which is termed one iteration. The iterative process continues until the total errors of all the samples or the change of the weights between two consecutive iterations is smaller than a predefined threshold value. The above algorithm is the basic BPNN training algorithm, which is also called a gradient descent algorithm. There are also other training algorithms for multi-layer feed-forward neural networks, such as Quasi-Newton and Levenberg-Marquardt algorithms.

B. Case Studies

BPNN is the most widely used neural network, having obtained successes in almost every modeling field, including pattern recognition for E-noses (Monge *et al.*, 2004; Yu and Wang, 2007; Tchoupo and Guiseppi-Elie, 2005; Kermani *et al*, 2005; Gardner *et al.*, 2000a; and Gardner *et al.*, 2000b).

- It was reported that BPNN has been successfully applied to E-noses for illness diagnoses (Gardner *et al.*, 2000b). In this study, both LDA and BPNN were applied to classify the bacteria: *S. aureus* and *E. coli* based on the headspace of their samples. LDA showed a clear overlap between the different classes, but BPNN provided a 100% identification rate for *S. aureus* and a 92% rate for *E. coli*. BPNN also yielded a good prediction of the bacterial growth phases, with the average rate being 80.7%.
- Kermani *et al.* (2005) also performed a comprehensive study of neural networks based E-nose pattern recognition systems. Two types of training algorithms, gradient descent algorithm and the Levenberg-Marquardt (LM) algorithm, were applied to train three-layer feed forward neural networks. Their pattern recognition performances were compared on a wide spectrum of samples: coffee, fragrance, environmental air, and cola. The training times of BP (gradient descent) algorithm for four tasks were 30.9 s (coffee), 123.0 s (fragrance), 13.2 s (environmental air), and 47.3 s (cola), respectively. Those of LM algorithm were 142.4 s (coffee), 48.3 s (fragrance), 2.4 s (environmental air), and 1215.6 s (cola), respectively. The recognition accuracies of BP algorithm were 68.4% (coffee), 72.2% (fragrance), 97.5% (environmental air), and 66.4 (cola), respectively, while those of LM algorithm were 98.4% (coffee), 86.9% (fragrance), 99.9% (environmental air), and 85.5% (cola). Clearly, the training times of BP algorithm were much shorter than those of LM algorithm in the tasks of coffee and cola classifications, but the accuracies of BP algorithm were all significantly poorer than those of LM algorithm except for the environmental air classification task. This study demonstrated the feasibility and reliability of neural network based E-nose pattern recognition strategy, although two training algorithms showed relatively salient difference in their recognition performance.

2.2.2. Radial Basis Function Neural Network (RBFN)

A. Principle

The learning process of BPNN is an iterative optimization process, so it could fall into local optima very easily during the learning process. However, the long learning time (i.e., convergence problem) is another possible major drawback of BPNN. These two potential problems often result in the failure of BPNN modeling. RBFN was first proposed in 1985 and is another type of widely used neural network.

The basic framework of RBFN is a radial function based interpolation problem. Supposing there are K p-dimensional learning samples $\{\mathbf{x}_i\}, i = 1,2,...,K$ and their corresponding outputs $\{\mathbf{y}_i\}$, RBFN is to seek an interpolation function as below:

$$f(\mathbf{x}_i) = \mathbf{y}_i$$

The typical structure of RBFN is also a three-layer BPNN. The hidden layer in RBFN performs a nonlinear transformation to transform the input space into a high dimensional transitional space by radial basis functions. The standard radial basis (transformation) function is Gaussian kernel function.

$$\varphi(\mathbf{x}, \mathbf{c}_j) = \exp(-\|\mathbf{x} - \mathbf{c}_j\|^2 / \sigma_j^2) \tag{18}$$

where, x is learning sample vector, c_j is the radial basis vector of the jth hidden node, σ_j is the Gaussian width of the jth hidden node.

The output layer produces linear weighting summation of all the hidden layer outputs.

$$f(\mathbf{x}) = \sum_{j=1}^{Q} \lambda_j \varphi(\mathbf{x}, \mathbf{c}_j) \tag{19}$$

where λ_j is the weight connecting the output node and the jth hidden node, and Q is the number of hidden nodes.

It is clear from eqs. (18,19) that the hidden layer in RBFN performs a fixed nonlinear Gaussian transformation based on the distance between input sample and the weight vector (radial basis vector) c_j of the jth hidden neuron. The parameters that need to be adjusted are only the weight between hidden and output layers λ_j. Usually, λ_j can be simply calculated by a regression model. Therefore as a one-step learning strategy, RBFN is able to avoid a long training time and local optima during the training process.

A successful RBFN model depends on proper radial basis vectors and the corresponding Gaussian widths. A conventional strategy to determine radial basis vector is K-means algorithm. This is an unsupervised learning algorithm and is able to group samples into a number of clusters based on the distances among samples. Thus, the cluster center vectors can be used as the radial basis vectors, as they reflect the space distribution of learning samples and theoretically are able to produce a good space transformation. Correspondingly, the Gaussian width for each radial basis center (i.e., K-means cluster center) can be considered as a function of the average distance between all the samples in this K-means cluster and the cluster center. In some cases, Gaussian widths are given the same value for all radial basis centers.

B. Cases Studies

- RBFN was successfully applied for bacteria classification (Dutta *et al.*, 2006). It was a six-class problem, consisting of *Staphylococcus aureus*, *Haemophilus influenzae*, *Escherichia coli*, *Pseudomonas aeruginosa*, *Moraxella catarrhalis*, and *Streptococcus pneumoniae*. The investigators increased the challenge of the problem by adding Gaussian noise to their sensor data. This operation lead to the occurrence of stochastic resonance (SR) in the data system and further resulted in an enhancement of the response of the E-nose system to the bacteria headspace signals. Three types of neural networks: multi-layer perceptron, probabilistic neural network and RBFN were employed for this classification. RBFN with the aid of the SR theory yielded a 98% accuracy in classifying six types of bacteria samples. In addition to proving the recognition effect of RBFN, this study suggested a promising data pre-processing method for pattern recognition.
- In a fragrance discrimination study (Branca *et al*, 2003), an expert group consisting of 20 trained panelists, GC/MS instruments, and radial basis function neural networks (RBFN) were employed to discriminate the fragrance with or without mangone (perfume note) at various diluted concentrations. Less than 40% panelists could identify the presence of the perfume note for concentration 0.01% or lower (w/w). The GC/MS instruments also performed poorly at these low concentrations. However, RBFN showed 100% recognition accuracy, even for the more diluted samples.

2.2.3. Time Delayed Neural Networks (TDNN)

A. Principle

In many cases, an E-nose sensor operates in real time and produces a time series of signals to represent a sample or a sample feature. The time series data provide implicit information about the trend of signal changes, and it becomes important to make full use of this information during pattern recognition. The neural network to be discussed in this section, TDNN, is a special neural network that works on input signals over time steps. TDNN was first successfully applied to speech (time series signals) recognition (Waibel *et al.*, 1989). Recently, TDNN has also been successfully applied to E-nose based classifiers (Zhang *et al.*, 2003, Zhang *et al.*, 2005).

The structure of a conventional TDNN is also a multi-layer forward neural network. The essential difference that distinguishes TDNN from BPNN and RBFN is the inputs of each TDNN neuron are composed of a number of signals over time steps. Figure 7 shows the input composition of each neuron. The training of TDNN is an iterative optimization process, so the error back-propagation algorithm is also suitable for TDNN. Employing time series information, TDNN has been proved to have a better cognition effect on time series samples than conventional BPNN and RBFN.

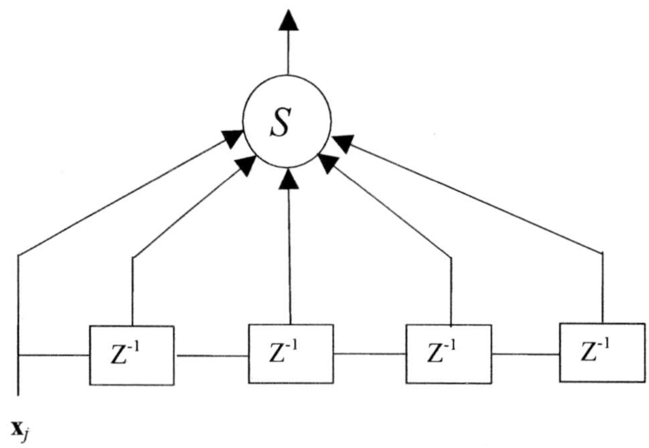

Figure 7. Single neuron structure of TDNN.

B. Case Studies

- TDNN was reported to have a successful application to the classification of four types of ground spices: basil, cardamom, pepper and turmeric (Zhang *et al.*, 2003). In this study, spice samples were measured using 12 polymer sensors each of which produced time series signals. Dsicriminant function analysis (DFA) and BPNN were also employed to compare with TDNN. For each sample, DFA and BPNN analyzed the instantaneous sensor data at the 4th min while TDNN analyzed the time series data collected every second up to the 4th min. The classification result for TDNN of a sample was based on the average of the neural network outputs over time. DFA and MLP correctly recognized 59.4 and 62.5% of the spices, respectively while TDNN correctly recognized all (100%) spices. In addition, this study also indicated that TDNN using time series data would outperform the methods using instantaneous data only when there is a significant difference among the time related structure of different samples. That is to say, sensor should provide the samples with different time series structures.
- TDNN was also successfully applied to quantify ingredient percentage of spice mixture using E-noses (Zhang *et al.*, 2005). The spice mixtures were composed of various fractions of three pure spices: basil, cinnamon and garlic. TDNN and other two types of neural networks, multilayer perceptron (MLP) and PCA-MLP, were applied to estimate the compositions of three pure spices in the spice mixtures. The E-nose systems in this study consisted of 12 polymer sensors to collect time series signals. All the three types of neural networks showed good estimation results. The mean square errors (MSE, in terms of the prediction error of the pure spice factions) for three types of neural networks were: 0.0051 (MLP), 0.0041 (PCA-MLP), and 0.0035 (TDNN). Again, employing the sensor data over each time step, TDNN showed a better modeling effect on the time series data.

2.2.4. Support Vector Machine (SVM)

A. Principle

SVM was proposed in the 1960's, but it was not studied extensively until the 1990's. Basically, SVM is a statistical method, but as we will discuss later, SVM has the similar structure to multi-layer feed-forward neural networks, so in many cases it is categorized into the neural network field.

SVM is becoming a novel and effective pattern recognition method for various fields including E-nose systems (Qian *et al.*, 2006; Gualdron *et al*, 2007; Pardo and Sberveglieri, 2005). Compared with statistical methods, SVM does not need *a priori* known model structure, which is the reason why SVM can be considered as a type of neural networks. Meanwhile, SVM is different from conventional neural networks. Its structure is determined by learning samples, unlike conventional feed-forward neural networks that need to define the number of hidden neurons before training. As a flexible tool, SVM has only recently been applied for E-nose systems, so it may be necessary to provide a relatively detailed description of its operation principle.

Again, let us use a two-class linear separation problem (Figure 8) for discussion. Suppose the linearly separable samples are $\{\mathbf{x}_i, y_i\}$, x is a *p*-dimensional vector, and $y_i = \{+1, -1\}$. The linear discrimination function for these two classes is:

$$f(\mathbf{x}_i) = \mathbf{w}^T \mathbf{x}_i + b = 0 \tag{20}$$

where, w is the weight vector and b is the threshold value.

It can be derived that the margin between two classes is $2/\|\mathbf{w}\|$ (Kecman, 2001). A super plane that can maximize the margin is the optimal separation plane. This optimization problem can be expressed as

$$\min : \Phi(\mathbf{w}) = \frac{1}{2}\|\mathbf{w}\|^2 = \frac{1}{2}\mathbf{w}^T \mathbf{w} \tag{21}$$

Introducing a Lagrange factor yields

$$\min : L(\mathbf{w}, b, \alpha) = \frac{1}{2}\mathbf{w}^T \mathbf{w} - \sum_i \alpha_i \{y_i[\mathbf{w}\mathbf{x}_i + b] - 1\} \tag{22}$$

where, α_i are Lagrange factors. The optimal solution is $\mathbf{w}^{opt} = \sum_i \alpha_i^{opt} y_i \mathbf{x}_i$. b^{opt} can be easily calculated by any support vector using $\mathbf{w}^{opt}$. Support vector are the samples whose $\alpha_i \neq 0$. Usually the α_i values for most of samples are zero, indicating they are not useful at all for pattern recognition. The black circles and black triangles in Figure 8 are the support vectors for these two classes.

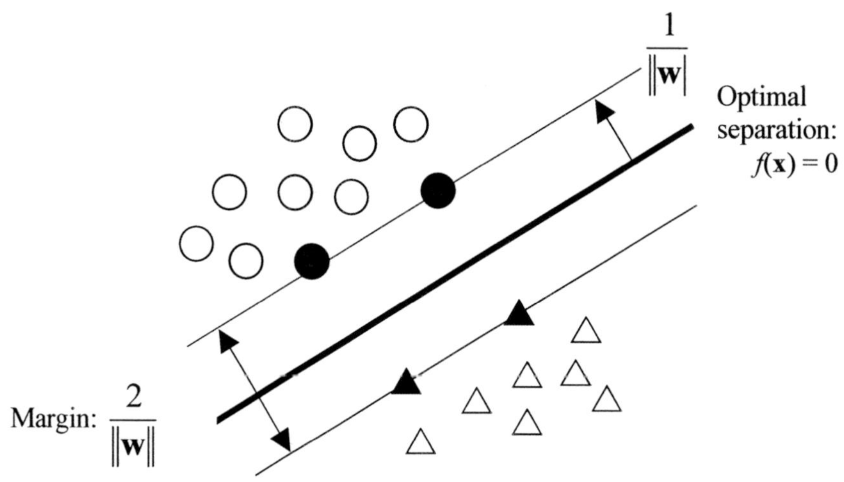

Figure 8. Support vectors based class separation.

The principle for linear separation can be easily extended to nonlinear systems. The framework of a SVM for a nonlinear system can be understood as: (1) to transform original input space to a multidimensional transitional space, and (2) to find an optimal super separation plane for the transitional space. The Gaussian kernel function is widely used as the transformation function. Clearly, the construction of a SVM for nonlinear classification is quite similar to that of RBFN. The major difference between them is that support vectors are selected from existing samples by using an optimization process while radial basis vector in RBFN are usually determined before network learning.

B. Case Studies

- One reported application of SVM in E-nose systems was the vapor detection and classification for environmental monitoring of spacecraft (Qian *et al.*, 2006). The E-nose system consisted of 32 sensors. SVM was first used for detect gas (vapor) in the spacecraft and had an 85% gas detection rate. Then, SVM was applied for the classification of 10 different single gas and 5 gas mixtures and achieve a 65.5% rate with the aid of the sensor selection technique.
- SVM has also obtained satisfactory results in the classification of food samples for a GC/MS based E-nose system (Gualdron *et al.*, 2007). The classification systems consisted of 11 classes of Spanish Ilbeian hams. Each class had roughly 10 samples. Sequential forward selection technique was applied to select important variables from the mass spectrometry m/z range. In this study, 11 SVM models (one-against-all) were created for 11 classes. Finally, with a variable selection technique, SVM had an accuracy of 90.3% for this 11-class problem.

2.2.5. Probability Neural Networks (PNN)

A. Principle

As an implementation of probability density function estimation and Baye's rule, the PNN was introduced for pattern recognition by Specht in 1988 (Specht, 1988). As discussed in the section of maximum probability rule, a Baye's classifier for an *N*-class problem can be briefly expressed as:

$$\mathbf{x} = \arg\max_h \left[P(\mathbf{x}/h)P(h)\right] \qquad (h = 1,2,\ldots,N)$$

where, $P(h)$ is the priori probability of class *h* and $P(\mathbf{x}/h)$ is the conditional probability of x given *h*. A standard estimator of $P(\mathbf{x}/h)$ is a Gaussian kernel function.

Essentially, PNN has the same principle as Baye's classifier. As shown in Figure 9, a standard PNN (Tian *et al.*, 1999) is composed of input layer, hidden layer, and output layer. The input layer receives the feature vectors, the hidden layer consists of all the training samples (but they are arranged into their known classes), and the output layer performs a summation process.

The output of the response of the *j*th training sample from class *h* ($\mathbf{x}_h^{(j)}$) to a *p*-dimensional input sample x can be expressed by a kernel function:

$$f(\mathbf{x}, \mathbf{x}_h^{(j)}, \sigma) = \frac{1}{K_h (2\pi)^{p/2} \sigma^p} \exp\left[-(\mathbf{x} - \mathbf{x}_h^{(j)})^T (\mathbf{x} - \mathbf{x}_h^{(j)})\right] \qquad (23)$$

where K_h is the number of the training samples from class *h*. The output neuron sums the responses of all the training samples from class *h* and uses it as the estimator of $P(\mathbf{x}/h)$. In some cases, considering the huge space required to store all the training samples, the mean vector or a number of sub-cluster center vectors of class *h* can be used in eq. (23).

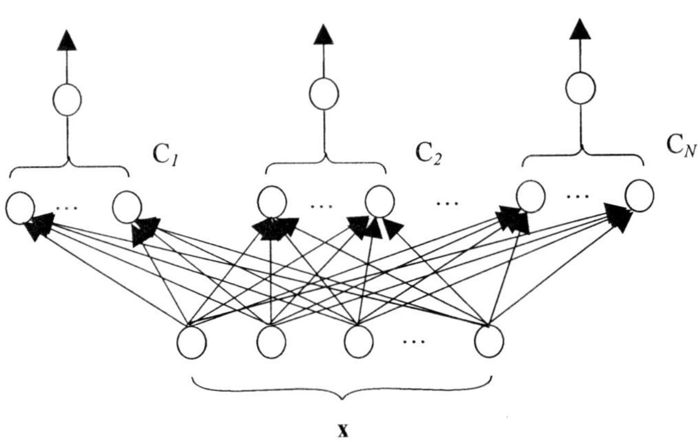

Figure 9. Structure of PNN.

Clearly, PNN can be considered as a neural network implementation of Baye's rule and a special type of RBFN as well.

B. Case Studies

PNN were reported to have good applications to E-nose pattern recognition and was often used to compare with RBFN due to their similarity in terms of structure and principle.

- PNN was also applied for the E-nose based identification of bacterial infections in hospital environments (Dutta *et al*., 2005). Similar to the case study reported in section 2.1.4, the identification system was composed of three types of bacterial, methicillin-resistant *S. aureus* (MRSA), methicillinsusceptible *S. aureus* (MSSA), and coagulase-negative *Staphylococci* (C-NS). The difference between them was the size of sample set. In this study, each type of bacterial infection had 50 samples that were collected from 50 in-hospital patients. PCA were used for data pre-processing first, and then three types of supervised neural networks, MLP, PNN, and RBFN were respectively applied to this classification task. The best classification result for PNN was 96.0% while that of RBFN was 99.7%.
- Another reported PNN application actually has been mentioned in the case study for RBFN in section 2.2.2. In that study (Dutta *et al*., 2006), PNN was also used to classify six types of bacteria: *Staphylococcus aureus*, *Haemophilus influenzae*, *Escherichia coli*, *Pseudomonas aeruginosa*, *Moraxella catarrhalis*, and *Streptococcus pneumoniae*. The accuracy of PNN for this six-class problem was 94%, a little lower than that of RBFN, 98%.

2.2.6. Fuzzy ARTMAP

A. Principle

Adaptive resonance theory neural networks have been successfully applied for cluster analysis, because of its stability and plasticity. One significant advantage of ART neural network is that adding a new training sample will not result in a wholly re-clustering process, so this type of neural networks are widely used for online data analysis. There are three major types of ART neural network, namely ART-1, ART-2, and ART-3. ART-1 processes (0,1) binary inputs while the other two can process analogue signals. This may be particularly helpful for E-nose systems.

The fuzzy ARTMAP to be discussed in this section consist of two Fuzzy ART-1s. Unlike ART-1, fuzzy ART-1 can process analogue patterns. Fuzzy set theory is used to calculate the similarity between an existing center and a new coming pattern to determine if there is a resonance between them. Fuzzy ARTMAP is an extension of ART and have been applied for E-nose pattern recognition (Vinaixa *et al*., 2005a; Vinaixa *et al*., 2004; Llobet *et al*., 2007; Gardner *et al*., 2000a). A brief description of the supervised learning process for fuzzy ARTMAP will be given below.

A fuzzy ARTMAP system is composed of two fuzzy ART modules (ARTa and ARTb) which are linked by an internal associate memory module. When the prediction of ARTa is not confirmed in ARTb, the associate memory field intrigues the match tracking rule so that the vigilance parameter in ARTa increases and category structure is reorganized. This way, the predictive error will not be repeated on subsequent training patterns.

B. Case Studies

- Fuzzy ARTMAP showed a good classification result in a study of qualitative assessment of food spoilage using the MS E-nose (Vinaixa, *et al.*, 2005b). The food was processed by four different rancidity accelerating treatments and grouped into four classes: fresh, rancidity A, B, and C. Each type contributed 9 samples for this 4-class problem. Each sample was analyzed by GC/MS and represented by a response spectrum obtained by averaging mass spectra along the detected unretained peaks. The MS range was m/z: 35-200, and the whole sample set was a 36 (samples) x 166 (m/z) matrix. First, the m/z fragments whose intensity did not have significant changes for the 36 measurements were excluded from analysis, and the m/z dimension decreased from 166 to 78. Then, PCA was employed for further dimension reduction and the first 10 PCs were retained for analysis. Finally a fuzzy ARTMAP was applied to create a classifier for this system. Thirty six samples were randomly reorganized to form 4 training-validating pairs, each of which had 28 training samples and 8 validating samples. The training process was based on "leave-one-out" strategy. Seventy eight m/z variables and their 10 PCs were respectively used as classification system input. The average (over four training-validating pairs) success rates in training and validating for fuzzy ARTMAP without PCA pre-processing were 87% and 88% while those for the fuzzy ARTMAP with PCA pre-processing were 96% and 93%. These results demonstrated that (1) fuzzy ARTMAP was an effective pattern recognition approach for the E-nose system, and (2) PCA can be a powerful data pre-processing method for pattern recognition.

2.3. Other Machine Learning Methods

This section will be mainly focused on some machine learning methods that are used as complementary or optimization tools to improve the performance of supervised learning algorithms for E-noses. Once a pattern recognition problem turns out to be a function map problem, optimization becomes the key issue for obtaining the best classifier parameters. Considering the drawbacks of conventional optimization tools, genetic algorithms (GA) or simulated annealing (SA) have been increasingly used for optimization processes. In addition, this section will also briefly introduce some unsupervised cluster analysis algorithms, as they either provide a visible clustering result of a sample set or work as a pre-processing method for supervised learning algorithms.

2.3.1. Genetic Algorithms (GA)

A. Principle

Most of pattern recognition problems eventually turn into optimization problems, so finding a robust and accurate optimization tool is a key issue for E-nose based pattern recognition. Genetic algorithms (GA) are a very effective optimization tool, based on global random searching strategy (Goldberg, 1989). The searching mechanism of GAs is a simulation of a biological nature evolution process. Unlike conventional and canonical optimization methods, GAs start the optimal solution search with a group of solution candidates not simply an initial point and do not require the optimization function to be continuous and derivable. These advantages make GAs become a powerful optimization tool especially for those complex, high dimensional and implicit optimization functions.

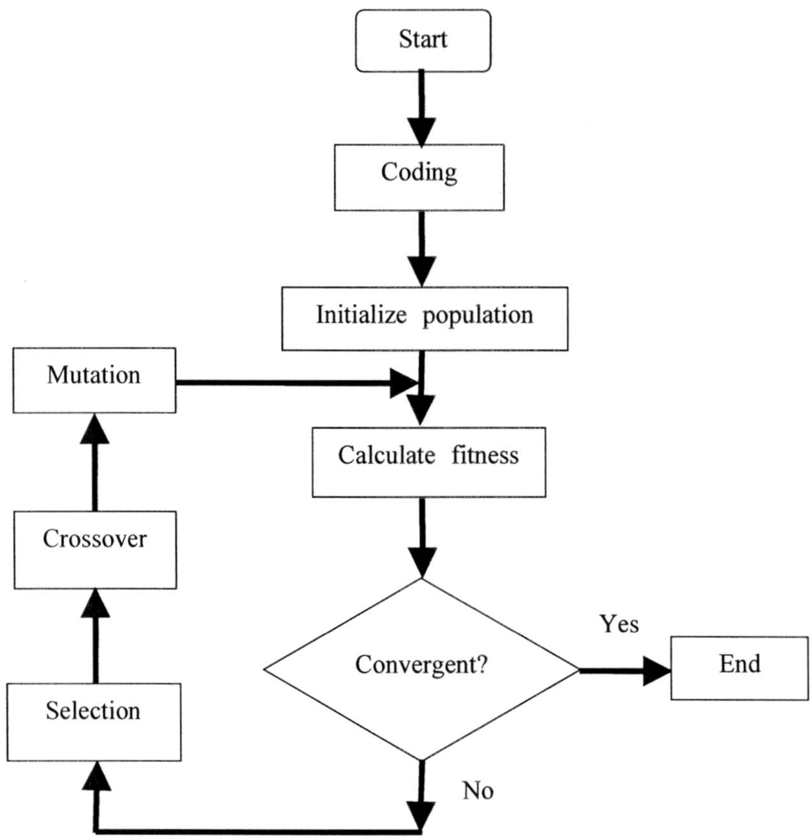

Figure 10. Flowchart of genetic algorithms.

GAs have two basic elements: solution expression mode (coding) and fitness function. There are two type of coding: binary and real. In GAs, a solution candidate is called a chromosome. In binary coding, each chromosome is expressed by a binary code string like [1101, 1001, 0011, ...], while in real coding, each chromosome is directly expressed by real numbers, like [v1, v2, v3, ...]. Fitness function is used to compare the property of each solution candidate. Usually a fitness function is the original objective function itself or a minor transformation (e.g., reciprocal) of the original objective function. Figure 10 shows the flowchart of GAs.

GAs consist of three major operators: selection, crossover, and mutation.

Selection: The algorithm selects candidates for next generation based on their fitness values. Roulette Wheel is widely used for selection. In general, a candidate with higher fitness may have relatively higher probability to be selected.
Crossover: In this step, two candidates will be used to generate their two son candidate.
Mutation: In this step, a randomly selected gene (variable) or a whole chromosome (solution vector) will be given a random modification within permitted variable ranges.

GAs continue performing these three operator until the convergence is reached. In many cases, a feasible convergent criterion is a predefined maximal iteration number. GAs are being increasingly used for various E-nose systems (Shnayderman *et al.*, 2005; Krebs *et al.*, 2006; Petricoin *et al.*, 2002; Eiceman *et al.*, 2006; Vinaixa *et al.*, 2005b).

B. Case Studies

- GAs were successfully applied for E-nose bacterial identification (Shnayderman *et al.*, 2005). In this study, the micromachined differential mobility spectrometer (microDMx) was used to analyze the headspace of four types of bacteria: *Escherichia coli*, *Bacillus subtilis*, *Bacillus thuringiensis*, and *Mycobacterium smegmatis*. The headspace gas spectra measured by the microDMx were investigated by pattern discovery/recognition algorithms to testify the feasibility to distinguish multiple species of bacteria *in vitro*. The entire data analysis section was divided into two parts: pattern discovery and pattern cognition. Genetic algorithms were employed in the pattern discovery part and were integrated with a lead cluster mapping to select a small subset of key features that were able to separate different classes. The highest overall validation accuracy of this study (84.2%) was achieved by a model with 11 features.
- The similar pattern discovery strategy based on genetic algorithm and a lead cluster mapping was also successfully used for species-species detection of *Bacillus* spores (Krebs *et al.*, 2006) and ovarian cancer identification (Petricoin *et al.*, 2002). In the study of *Bacillus* spore detection, pyrolysis-differential mobility spectrometry was used to analyze the samples. The pattern discovery strategy successfully distinguished *B. subtilis* from *B. thuringiensis* and *B. cereus*, yielding a high validation accuracy (>90%) to both *B. thuringiensis* and *B. cereus*. This study provided a promising direction for constructing an E-nose based bio-warfare agent detection system.
- Another relatively detailed description of GA application was reported in the study (Vinaixa *et al.*, 2005b) which has been used to illustrate fuzzy ARTMAP in section 2.2.6. In fact, another task of that study was to use PLS model to predict Rancimat and acid degree value (ADV) from the m/z values of the food samples. As introduced in section 2.2.6, 78 m/z variables were retained for analysis, but there could still be

some useless even noisy information among these 78 variables. Therefore GAs were employed to select "good" variables from the 78 m/z responses for PLS modeling. The coding method was to use a 78-dimensional binary vector (0,1) to represent if an m/z variable is useful ("1") or not ("0"). The fitness function was based on the prediction error of the PLS model. The lower the error, the higher the fitness. Finally, GAs selected 14 m/z variables for each single prediction model for Rancimat and ADV and selected 15 m/z variables for the simultaneous prediction for both Rancimat and ADV. The correlations of the predicted and measured values for Rancimat and ADV were up to 0.98 and 0.97. Although this application was for correlation, the method in this application could be easily used for pattern recognition, as pattern recognition can be considered as a specific type of correlation.

2.3.2. Simulated Annealing (SA)

A. Principle

Simulated annealing (SA) is another global searching method motivated by the annealing process in metallurgy. In an optimum searching process, SA algorithms replace the solution at each step with a random neighbor solution, based on the current "annealing" temperature (T). This temperature gradually decreases during the process, so the solutions in earlier steps may have larger disturbances to change their status. This replacement can help prevent searching process falling into local optima. The key for SA is to establish a proper "annealing" temperature (T) decrease function. In some cases, SA may require overlong time to follow a fine-tuned decreasing process of T.

B. Case Studies

- A reported successful application of SA to E-nose systems was to select variables for pattern recognition. In the study of the classification of 11 classes of Ilberian ham (Llobet *et al.*, 2007), SA was used to select "good" m/z variables from the m/z range (m/z: 45-250) for fuzzy ARTMAP. The detailed description of the samples has been given in section 2.2.4. The cost function for SA was the prediction error of the fuzzy ARTMAP. Nearly 80% of the original m/z variables were removed form analysis by SA and finally the fuzzy ARTMAP had a 97.2% accuracy of this 11-class problem. This result was higher than the best SVM result for this problem (90.3%).

2.3.3. K-means

A. Principle

K-means is an unsupervised learning based cluster analysis method. The basic concept of this algorithm is to minimize the sum of the distances between all the samples to their cluster centers (i.e., total intra-cluster variance). Assuming n samples need to be grouped into k clusters, the major steps of this algorithm are:

(1) Randomly select k sample from the sample set as initial cluster centers, $\mathbf{z}_1^1, \mathbf{z}_2^1, \ldots, \mathbf{z}_k^1$.

(2) In the qth iteration, for a sample x, if $\left|\mathbf{x}-\mathbf{z}_j^q\right| < \left|\mathbf{x}-\mathbf{z}_i^q\right|, (i=1,2,\cdots,n, i \neq j)$, then x is grouped into class j. Similarly, group all the samples into their closest classes.

(3) Update cluster center

$$\mathbf{z}_j^{q+1} = \frac{1}{n_j} \sum_{\mathbf{x} \in S_j^q} \mathbf{x} \quad (j=1,2,\cdots,k)$$

where, S_j^q is the set of the samples belonging to class j in the qth iteration, and n_j is the number of the samples in S_j^q.

If $\mathbf{z}_j^{q+1} = \mathbf{z}_j^q (j=1,2,\cdots,k)$, training process ends. Otherwise go back to step (2).

K-means is a simple, but effective, cluster analysis method and has been widely used as data compression and feature extraction tools for pattern recognition algorithms applied to E-nose systems. For example, K-means is a practical method to determine radial basis vectors for RBFN and center vectors for PNN. One drawback of K-means is that the global convergence of this algorithm largely depends on the initial sample set. Another possible problem is that may have a skewed clustering result, if the cluster number estimate is incorrect.

In the reports of the E-nose based pattern recognition we have recently collected, as clustering methods, the K-means method and the self-organization map to be discussed below were mainly used to help constructing neural networks or provide a primary visible classification result. There are few independent usages of these two methods in the E-nose systems, particularly in the supervised learning based pattern recognition studies. Therefore, the Case Studies section will not be given to these two methods.

2.3.4. Self-organization Map (SOM)

A. Principle

The self-organization map (SOM) is an unsupervised learning neural network. The most significant feature of SOM is the capability to project a high-dimensional space onto a two-dimensional space while preserving the topology property of high-dimensional samples. Therefore SOM can be used for both dimension reduction and similarity display and has been employed for E-nose construction (Dutta *et al.*, 2005, Lamagna *et al.*, 2004).

Figure 12 shows the standard structure of SOM. SOM is a one-layer neural network. The learning algorithm of SOM is a competitive algorithm and can be summarized as three major steps: (1) Find the weight vector which is closest to the input training vector and the corresponding neuron can be called winner neuron, (2) Adjust this weight vector towards the input vector, and (3) Adjust the weight vectors of the neurons within the neighbor area of the winner neuron. Therefore, the success of SOM depends on the learning rate for weight adjusting and the neighbor area that decreases during the training process. These three steps fully reflect the mechanism of signal transferring among brain neurons.

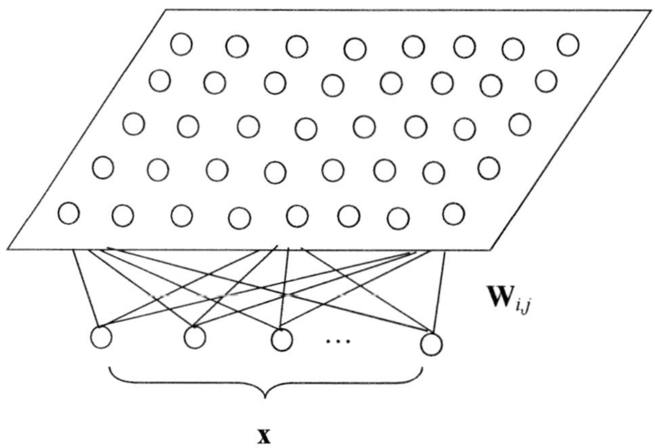

Figure 12. Structure of SOM.

3. Conclusion

The most popular application field of E-nose systems to date has been in the food industries, clinical diagnostics, and environmental monitoring. In terms of pattern distribution, pattern recognition methods can be divided into linear and nonlinear approaches. ANN obviously is a powerful tool for nonlinear classification. Probability rule based methods are also capable to classify nonlinear distributed patterns. Linear discrimination analysis or other linear methods can be extended to nonlinear fields, if stepwise linear functions are applied. In general, an E-nose sensor system could consist of many sensors, and so data pre-processing (especially dimension reduction) must be the first step in many cases. PCA is the most widely used data pre-processing technique for dimension reduction, noise removal, and primary observation of pattern space distribution. In addition, there are other variable selection strategies that are able to directly select "good" or "distinct" variables to improve modeling effect. The Gaussian kernel function is often used as the basic structural element for SVM, RBFN, PNN, and some types of maximum probability classifiers. These methods can be categorized into kernel machine and their common parameters to be determined are basis vectors and Gaussian widths. In some cases, non-parametric methods or adaptive models could yield better results than conventional fixed parametric models. GA and SA can optimize model parameters and structures for high pattern recognition accuracy.

A good E-nose based classifier mainly depends on sensors for collecting signals (hardware) and algorithms for classification (software). In the aspect of classification algorithms, the following directions should be given sufficient concerns. In feature extraction, it is expected to extract the most representative and relevant information for pattern recognition. In many cases, E-nose sensors provide time series of signals to classifier, so digital signal processing and time series analysis methods will have more applications to feature extraction. In a recent report (Krebs *et al.*, 2005), auto-regression (AR) model was used for extract the features from the GC profiles of the samples. The AR coefficients showed a good visible separation among four types of bacteria.

Overall, there are four key issues that dominate algorithm choices for E-nose applications: training time, accuracy, robustness, and adaptivity. There may be a trade-off among these four indexes, which is a challenge for every pattern recognition researcher. Integration of methods with different properties is still a promising direction to improve pattern recognition accuracy. Artificial intelligence will continue working as a powerful machine learning method for E-nose based pattern recognition.

Acknowledgements

This publication was made possible due to generous research support during the preparation of this book chapter. The authors would like to acknowledge and thank the Army Research Office (W911NF-06-1-0272), the Defense Advanced Research Projects Agency Microsystems Technology Office (PM Dennis Polla), Lawrence Livermore National Laboratory (subcontract B563119 from W-7405-ENG-48), and grant number UL1 RR024146 from the National Center for Research Resources (NCRR), a component of the National Institutes of Health (NIH), and NIH Roadmap for Medical Research.

The contents of this book chapter are solely the responsibility of the authors and do not necessarily represent the official view of the funding agencies.

References

T. Aishima, 2004. Correlating sensory attributes to gas chromatography–mass spectrometry profiles and e-nose responses using partial least squares regression analysis. *Journal of Chromatography A* **1054**, 39–46.

A. C. Bastos, N. Magan, 2006. Potential of an electronic nose for the early detection and differentiation between Streptomyces in potable water. *Sensors and Actuators B* **116**, 151–155.

A. Z. Berna, J. Lammertyn, S. Saevels, C. D. Natale, B. M. Nicolai, 2004. Electronic nose systems to study shelf life and cultivar effect on tomato aroma profile. *Sensors and Actuators B* **97**, 324–333.

A. Branca, P. Simonian, M. Ferrante, E. Novas, R. Negri, 2003. Electronic nose based discrimination of a perfumery compound in a fragrance. *Sensors and Actuators B* **92**, 222–227.

S. Buratti, S. Benedettia, M. Scampicchio, E.C. Pangerod, 2004. Characterization and classification of Italian Barbera wines by using an electronic nose and an amperometric electronic tongue. *Analytica Chimica Acta* **525**, 133–139.

M. Carmona, J. Martinez, A. Zalacain, M. L. Rodriguez-Mendez, J. A. de Saja, G. L. Alonso, 2006. Analysis of saffron volatile fraction by TD–GC–MS and e-nose. *European Food Research and Technology* **223**: 96–101.

X. Chen, M. Cao, Y. Li, W. Hu, P. Wang, K. Ying, H. Pan, 2005. A study of an electronic nose for detection of lung cancer based on a virtual SAW gas sensors array and imaging recognition method. *Measurement Science and Technology* **16**, 1535–1546.

W. A. Collier, D. B. Baird, Z. A. Park-Ng, N. More, A. L. Hart, 2003. Discrimination among milks and cultured dairy products using screen-printed electrochemical arrays and an electronic nose. *Sensors and Actuators B* **92**, 232–239.

D. Cozzolino, H. E. Smyth, W. Cynkar, R. G. Dambergs, M. Gishen, 2005. Usefulness of chemometrics and mass spectrometry-based electronic nose to classify Australian white wines by their varietal origin. *Talanta* **68**, 382–387.

T. H. Dodd, S. A. Hale, S. M. Blanchard, 2004. Electronic nose analysis of Tilapia storage. *Transactions of the ASAE* **47**(1), 135-140.

R. Dutta, D. Morgan, N. Baker, J. W. Gardner, E. L. Hines, 2005. Identification of Staphylococcus aureus infections in hospital environment: electronic nose based approach. *Sensors and Actuators B* **109**, 355–362.

R. Dutta, A. Das, N. G. Stocks , D. Morgan, 2006. Stochastic resonance-based electronic nose: A novel way to classify bacteria. *Sensors and Actuators B* **115**, 17–27.

R. Dutta, R. Dutta, 2006. "Maximum probability rule" based classification of MRSA infections in hospital environment: Using electronic nose. *Sensors and Actuators B* **120**, 156–165.

G. A. Eiceman, M. Wang, S. Prasad, H. Schmidt, F.K. Tadjimukhamedov, Barry K. Lavine, N. Mirjankar, 2006. Pattern recognition analysis of differential mobility spectra with classification by chemical family. *Analytica Chimica Acta* **579**,1–10.

R. A. Fisher, 1936. The Use of Multiple Measurements in Taxonomic Problems. *Annals of Eugenics* **7**,179-188.

J. W. Gardner, H. W. Shin, E. L. Hines, C. S. Dow, 2000a. An electronic nose system for monitoring the quality of potable water. *Sensors and Actuators B* **69**, 336–341.

J. W. Gardner, H. W. Shin, E. L. Hines, 2000b. An electronic nose system to diagnose illness. *Sensors and Actuators B* **70**,19–24.

D. E. Goldberg, 1989. *Genetic Algorithms in Search, Optimization and Machine Learning*, Kluwer Academic Publishers, Boston, MA.

O. Gualdron, J. Brezmes, E. Llobet, A. Amari, X. Vilanova, B. Bouchikhi, X. Correig, 2007. Variable selection for support vector machine based multisensor systems. *Sensors and Actuators B* **122**, 259-268.

S. Hamilton, M.J. Hepher, J. Sommerville, 2005. Polypyrrole materials for detection and discrimination of volatile organic compounds. *Sensors and Actuators B* **107**, 424–432.

A. Hoskuldsson, 1988. PLS regression methods. *Journal of Chemometrics* **2,** 211-228.

V. Kecman, 2001. *Learning and Soft Computing, Support Vector Machines, Neural Networks and Fuzzy Logic Models*, The MIT Press, Cambridge, MA, USA.

B. G. Kermani, S. S. Schiffman, H. T. Nagle, 2005. Performance of the Levenberg–Marquardt neural network training method in electronic nose applications. *Sensors and Actuators B* **110**, 13–22.

M. D. Krebs, R. D. Tingley, J. E. Zeskind, J.-M. Kang, M. E. Holmboe, C. E. Davis, 2005. Autoregressive modeling of analytical sensor data can yield classifiers in the predictor coefficient parameter space. *Bioinformatics* **21** (8), 1325-1331.

M. D. Krebs, B. Mansfield, P. Yip, S. J. Cohen, A. L. Sonenshein, B. A. Hitt, C. E. Davis, 2006. Novel technology for rapid species-specific detection of Bacillus spores. *Biomolecular Engineering* **23**, 119–127.

M. Kuske, A.-C. Romain, J. Nicolas, 2005. Microbial volatile organic compounds as indicators of fungi. Can an electronic nose detect fungi in indoor environments? *Building and Environment* **40**, 824–831.

A. Lamagna, S. Reich, D. Rodriguez, N.N. Scoccola, 2004. Performance of an e-nose in hops classification. *Sensors and Actuators B* **102**, 278–283.

E. Llobet, O. Gualdrón, M. Vinaixa, N. El-Barbri, J. Brezmes, X. Vilanova, B. Bouchikhi, R. Gómez, J.A. Carrasco, X. Correig, 2007. Efficient feature selection for mass spectrometry based electronic nose applications. *Chemometrics and Intelligent Laboratory Systems* **85**, 253-261.

M. Lovino, M. F. Cardinal, D. B.V. Zubiri, D. L. Bernik, 2005. Electronic nose screening of ethanol release during sol–gel encapsulation A novel non-invasive method to test silica polymerization. *Biosensors and Bioelectronics* **21**, 857–862.

J. Lozano, J. P. Santos, M. Aleixandre, Is. Sayago, J. Gutiérrez, M. Carmen Horrillo, 2006. Identification of Typical Wine Aromas by Means of an Electronic Nose. *IEEE Sensors Journal* **6**(1), 173-178.

D. Luo, H. G. Hosseini, J. R. Stewart, 2004. Application of ANN with extracted parameters from an electronic nose in cigarette brand identification. *Sensors and Actuators B* **99**, 253–257.

M. P. Marti, O. Busto, J. Guasch, 2004. Application of a headspace mass spectrometry system to the differentiation and classification of wines according to their origin, variety and ageing. *Journal of Chromatography A* **1057**, 211–217.

J. R. Minkel, 2005. e-Nose Sniffs Out Cancer, and More. *The Scientist*, **19**(17), 36.

M. E. Monge, D. Bulone, D. Giacomazza, D. L. Bernik, R. M. Negri, 2004. Detection of flavour release from pectin gels using electronic noses. *Sensors and Actuators B* **101**, 28–38.

J. Nicolas, A.-C. Romain, V. Wiertz, J. Maternova, Ph. Andre, 2000. Using the classification model of an electronic nose to assign unknown malodours to environmental sources and to monitor them continuously. *Sensors and Actuators B* **69**, 366–371.

G. Onkal-Engin, I. Demir, S. N. Engin, 2005a. Determination of the relationship between sewage odour and BOD by neural networks. *Environmental Modelling & Software* **20**, 843-850.

G Önkal-Engin, I. Demir, S. N. Engin, 2005b. e-NOSE Response Classification of Sewage Odors by Neural Networks and Fuzzy Clustering. *Advances in Natural Computation-ICNC 2005, LNCS* **3611**, 648 – 651.

M. Pardo, G. Sberveglieri, 2005. Classification of electronic nose data with support vector machines. *Sensors and Actuators B* **107**, 730–737.

A. Pavlou, A.P.F. Turner , N. Magan, 2002. Recognition of anaerobic bacterial isolates in vitro using electronic nose technology. *Letters in Applied Microbiology* **35**, 366–369.

E. F Petricoin III, A. M. Ardekani, B. A. Hitt, P. J. Levine, V. A. Fusaro, S. M. Steinberg, G. B. Mills, C. Simone, D. A. Fishman, E. C. Kohn, L. A. Liotta, 2002. Use of proteomic patterns in serum to identify ovarian cancer. *The Lancet* **359**, 572–77.

S. Prasad, H. Schmidt, P. Lampen, M. Wang, R. Guth, J.V. Rao, G. B. Smith, G. A. Eiceman, 2006. Analysis of bacterial strains with pyrolysis-gas chromatography/differential mobility spectrometry. *Analyst* **131**, 1216–1225.

D. S. Presicce, A. Forleo, A.M. Taurino, M. Zuppa, P. Siciliano, B. Laddomada, A. Logrieco, A. Visconti, 2006. Response evaluation of an E-nose towards contaminated wheat by Fusarium poae fungi. *Sensors and Actuators B* **118**, 433–438.

T. Qian, X. Li, B. Ayhan, R. Xu, C. Kwan, T. Griffin, 2006. Application of Support Vector Machines to Vapor Detection and Classification for Environmental Monitoring of Spacecraft. *Advances in Neural Networks-ISNN 2006, LNCS* **3973**, 1216 – 1222.

S. M. Rocha, P. Coutinho, A. Barros, I. Delgadillo, M. A. Coimbra, 2006. Rapid tool for distinction of wines based on the global volatile signature. *Journal of Chromatography A* **1114**, 188–197.

A. C. Romain, D. Godefroid, M. Kuske, J. Nicolas, 2005. Monitoring the exhaust air of a compost pile as a process variable with an e-nose. *Sensors and Actuators B* **106**, 29–35.

E. Scorsone, A. M. Pisanelli, K. C. Persaud, 2006. Development of an electronic nose for fire detection. *Sensors and Actuators B* **116**, 55–61.

M. Shnayderman, B. Mansfield, P. Yip, H. A. Clark, M. D. Krebs, S. J. Cohen, J. E. Zeskind, E. T. Ryan, H. L. Dorkin, M. V. Callahan, T. O. Stair, J. A. Gelfand, C. J. Gill, B. Hitt, C. E. Davis, 2005. Species-Specific Bacteria Identification Using Differential Mobility Spectrometry and Bioinformatics Pattern Recognition. *Analytical Chemistry* **77**, 5930-5937.

C. O. Sim, M. N. Ahmad, Z. Ismail, A. R. Othman, N. A. M. Noor, E. M. Zaihidee, 2003. Chemometric Classification of Herb – *Orthosiphon stamineus* According to Its Geographical Origin Using Virtual Chemical Sensor Based Upon Fast GC. *Sensors* **3**, 458-471.

D. Specht, 1988. Probabilistic neural networks for classification, mapping, or associative memory. *Proceeding of the IEEE International Conference on Neural Networks* **1**, 525-532.

S. Strathmann, S. Pastorelli, C. Simoneau, 2005. Investigation of the interaction of active packaging material with food aroma compounds. *Sensors and Actuators B* **106**, 83–87.

G. N. Tchoupo, A. Guiseppi-Elie, 2005. On pattern recognition dependency of desorption heat, activation energy, and temperature of polymer-based VOC sensors for the electronic NOSE. *Sensors and Actuators B* **110**, 81–88.

B. Tian, M. A. Shaikh, M. R. Azimi-Sadjadi, T. H. V. Haar, D. L. Reinke, 1999. A Study of Cloud Classification with Neural Networks Using Spectral and Textural Features. *IEEE Transactions on Neural Networks* **10**(1), 138-151.

L. Torsi, A. Dodabalapur, L. Sabbatini, P.G. Zambonin, 2000. Multi-parameter gas sensors based on organic thin-film-transistors. *Sensors and Actuators B* **67**, 312–316.

J. Trihaas, L. Vognsen, P.V. Nielsen, 2005a. Electronic nose: New tool in modeling the ripening of Danish blue cheese. *International Dairy Journal* **15**, 679–691.

J. Trihass, T. V. D. Tempel, P. V. Nielsen, 2005b. Electronic Nose Technology in Quality Assessment: Predicting Volatile Composition of Danish Blue Cheese During Ripening. *Journal of Food Science* **70**(6), E392-E400.

J. S. Vestergaard, J.-E. Haugen , D. V. Byrne, 2006. Application of an electronic nose for measurements of boar taint in entire male pigs. *Meat Science* **74**, 564–577.

M. Vinaixa, S. Marlan, J. S. Brezmes, E. Llobet, X. Vilanova, X. Correig, A. Ramos, V. Sanchis, 2004. Early Detection of Fungal Growth in Bakery Products by Use of an Electronic Nose Based on Mass Spectrometry. *Journal of Agricultural and Food Chemisty* **52**, 6068-6074.

M. Vinaixa , A. Vergara , C. Duran , E. Llobet , C. Badia , J. Brezmes , X. Vilanova , X. Correig, 2005a. Fast detection of rancidity in potato crisps using e-noses based on mass spectrometry or gas sensors. *Sensors and Actuators B* **106**, 67–75.

M. Vinaixa, E. Llobet, J. Brezmes, X. Vilanova, X. Correig, 2005b. A fuzzy ARTMAP- and PLS-based MS e-nose for the qualitative and quantitative assessment of rancidity in crisps. *Sensors and Actuators B* **106**, 677–686.

A. Waibel, T. Hanazawa, G. Hinton, K. Shikano, K. J. Lang, 1989. Phoneme Recognition Using Time-Delay Neural Networks. IEEE *Transactions on Acoustics, Speech, and Signal Processing* **37**(3), 328-339.

Y. S. Yang, S.-C. Ha, Y. S. Kim, 2005. A matched-profile method for simple and robust vapor recognition in electronic nose (E-nose) system. *Sensors and Actuators B* **106**, 263–270.

R. C. Young, W. J. Buttner, B. R. Linnell, R. Ramesham, 2003. Electronic nose for space program applications. *Sensors and Actuators B* **93**, 7–16.

H. Yu, J. Wang, 2007. Discrimination of LongJing green-tea grade by electronic nose. *Sensors and Actuators B* **122**, 134-140.

H. Zhang , M. Ö. Balaban , J. C. Principe, 2003. Improving pattern recognition of electronic nose data with time-delay neural networks. *Sensors and Actuators B* **96**, 385–389.

H. Zhang, M. Ö. Balaban, J. C. Principe, K. Portier, 2005. Quantification of Spice Mixture Compositions by Electronic Nose: Part I. Experimental Design and Data Analysis Using Neural Networks. *Journal of Food Science* **70**(4), E253-E258.

In: Pattern Recognition in Nanoscience...
Editor: Diane M. Perez, pp. 43-58
ISBN: 1-60021-715-X

Chapter 2

APPLICATION OF PATTERN RECOGNITION IN WASTEWATER TREATMENT PLANTS

C. Riverol* and V. Parsadie
Chemical Engineering Department, University of West Indies,
St. Augustine Campus, Trinidad,
Republic of Trinidad and Tobago

Abstract

The most important task in knowledge-based recognition procedure is the conversion of input process variables into features. The second is the translation of the classification logic expressed initially in natural linguistic form into a transparent recognition algorithm that yields from the structural variables membership to different classes. Pattern recognition is a potentially useful approach for interpreting the data generated by chemical processes. In contrast to expert system approaches to data interpretation where analysis proceeds deductively in a sequence of logical steps, pattern recognition involves inductive reasoning through generalization from a set of learned examples of process behavior. Pattern recognition may also complement the expert system approach by carrying out the low level assessment of process trends not easily captured in rules.

Wastewater processing by means of activated sludge is most widespread sewage treatment process. The activated sludge process (ASP) is with the further complication of operating in a widely changing environment where process conditions may be appreciably altered by immeasurable inputs.

This chapter introduces the field of pattern recognition to the wastewater treatment plant in a much broader scope. This chapter analyses the theory involved in modeling the bioreactor (the major process equipments involved) followed by a comparative study of the IAWQ state-of-the-art model ASM1 with ASM3. It can be used as an introduction to the field in that it touches on the main branches of this discipline. No attempt will be made to introduce the reader to the broad field of artificial intelligence. Suffice to say that pattern recognition is a subset of artificial intelligence. This approach allows modeling and simulating an activated sludge process using its dynamic characteristics.

1. Introduction

The activated sludge process (ASP) is a biological technique for treating wastewater. This process was developed in England from 1912 to 1914 by Ardern and Lockett [1]. By means of discontinuous aeration experiments, it was discovered that wastewater cannot be cleaned exclusively by aeration from the contained pollutants, but that it must be activated with sludge and/or the living organisms contained in it. In this process, the settled sewage is aerated in tanks, by compressed air or mechanical agitation, in the presence of 'activated sludge'- a flocculent settleable sludge consisting of cultures of micro-organisms which gradually accumulates in these conditions. The aeration of this mixed liquor is continued for several hours, during which the colloids become flocculated and the impurities in the sewage undergo biological oxidation. The sludge is then separated from the purified sewage by settlement in tanks, and a portion of it is returned to the aeration tanks to treat more sewage. The excess sludge which is continually accumulating during the process is withdrawn for disposal elsewhere.

2. The IAWQ Activated Sludge Models

In 1983, the International Association on Water Quality (IAWQ, formerly IAWPRC) formed a task group, which was to promote development and facilitate the application of, practical models for design and operation of biological wastewater treatment systems. The first goal was to review existing models and the second goal was to reach a consensus concerning the simplest mathematical model having the capability of realistically predicting the performance of single-sludge systems carrying out carbon oxidation, nitrification and denitrification. The final result was presented in 1987 by Henze [2] as the IAWQ Activated Sludge Model No. 1. Although the model has been extended since then, for example to incorporate more fractions of COD to accommodate new experimental observations [3], to describe growth and population dynamics of floc forming and filamentous bacteria [4] and to include new processes for describing enhanced biological phosphorous removal [5] the original model is the most widely used for describing WWTP all over the world.

2.1. The Activated Sludge Model 1, ASM1

ASM1 is presented in a matrix format in Table 1 according to [2].

Table 1. The ASM1 process matrix (Henze et al., 1987)

Component → *i*		1	2	3	4	5	6	7	8	9
j	Process ↓	S_I	S_S	X_I	X_S	$X_{B,H}$	$X_{B,A}$	X_P	S_O	S_{NO}
1	Aerobic growth of heterotrophs		$-\frac{1}{Y_H}$			1			$-\frac{1-Y_H}{Y_H}$	
2	Anoxic growth of heterotrophs		$-\frac{1}{Y_H}$			1				$-\frac{1-Y_H}{2.86Y_H}$
3	Aerobic growth of autotrophs						1		$-\frac{4.57}{Y_A}+1$	$\frac{1}{Y_A}$
4	'Decay' of heterotrophs				$1-f_P$	-1		f_P		
5	'Decay' of autotrophs				$1-f_P$		-1	f_P		
6	Ammonification of soluble organic nitrogen									
7	'Hydrolysis' of entrapped organics		1		-1					
8	'Hydrolysis' of entrapped organic nitrogen									
Observed Conversion Rates $[ML^{-3}T^{-1}]$		$r_i = \sum_j \nu_{ij}\rho_j$								
Stoichiometric Parameters: Heterotrophic yield: Y_H Autotrophic yield: Y_A Fraction of biomass yielding particulate products: f_P Mass N/Mass COD in biomass: i_{XB} Mass N/Mass COD in products from biomass: i_{XP}		Soluble inert organic matter $[M(COD)L^{-3}]$	Readily biodegradable substrate $[M(COD)L^{-3}]$	Particulate inert organic matter $[M(COD)L^{-3}]$	Slowly biodegradable substrate $[M(COD)L^{-3}]$	Active heterotrophic biomass $[M(COD)L^{-3}]$	Active autotrophic biomass $[M(COD)L^{-3}]$	Particulate products arising from biomass decay $[M(COD)L^{-3}]$	Oxygen (negative COD) $[M(-COD)L^{-3}]$	Nitrate and nitrite nitrogen $[M(N)L^{-3}]$

Table 1. (Continued)

10 S_{NH}	11 S_{ND}	12 X_{ND}	13 S_{ALK}	Process Rate, ρ_j [ML-3T-1]
$-i_{XB}$			$\frac{i_{XB}}{14}$	$\hat{\mu}_H\left(\frac{S_S}{K_S+S_S}\right)\left(\frac{S_O}{K_{O,H}+S_O}\right)X_{B,H}$
$-i_{XB}$			$\frac{1-Y_H}{14\cdot 2.86Y_H}$ $-\frac{i_{XB}}{14}$	$\hat{\mu}_H\left(\frac{S_S}{K_S+S_S}\right)\left(\frac{K_{O,H}}{K_{O,H}+S_O}\right)\cdot\left(\frac{S_{NO}}{K_{NO}+S_{NO}}\right)\eta_g X_{B,H}$
$-i_{XB}-\frac{1}{Y_A}$			$-\frac{i_{XB}}{14}-\frac{1}{7Y_A}$	$\hat{\mu}_A\left(\frac{S_{NH}}{K_{NH}+S_{NH}}\right)\left(\frac{S_O}{K_{O,A}+S_O}\right)X_{B,A}$
		$i_{XB}-f_P i_{XP}$		$b_H X_{B,H}$
		$i_{XB}-f_P i_{XP}$		$b_A X_{B,A}$
1	−1		$\frac{1}{14}$	$k_a S_{ND} X_{B,H}$
				$k_h\frac{X_S/X_{B,H}}{K_X+(X_S/X_{B,H})}\left[\left(\frac{S_O}{K_{O,H}+S_O}\right)+\eta_h\left(\frac{K_{O,H}}{K_{O,H}+S_O}\right)\left(\frac{S_{NO}}{K_{NO}+S_{NO}}\right)\right]X_{B,H}$
	1	−1		$\rho_7(X_{ND}/X_S)$
$r_i=\sum_j v_{ij}\rho_j$				
NH4+NH3 nitrogen [M(N)L-3]	Soluble biodegradable organic nitrogen [M(N)L-3]	Particulate biodegradable organic nitrogen [M(N)L-3]	Alkalinity – Molar units	Kinetic Parameters: Heterotrophic growth and decay: $\hat{\mu}_H, K_S, K_{O,H}, K_{NO}, b_H$ Autotrophic growth and decay: $\hat{\mu}_A, K_{NH}, K_{O,A}, b_A$ Correction factor for anoxic growth of heterotrophs: η_g Ammonification: k_a Hydrolysis: k_h, K_X Correction factor for anoxic hydrolysis: η_h

○ **COD Components in ASM1**

COD is selected as the most suitable parameter for defining the carbon substrates as it provides a link between electron equivalents in the organic substrate, the biomass and oxygen utilized. In ASM1 the COD is subdivided based on (1) solubility, (2) biodegradability (3) biodegradation rate and (4) viability (biomass):

The total COD, is divided into soluble (*S*) and particulate (*X*) components.

- The COD is further subdivided into non-biodegradable organic matter and biodegradable matter. The non-biodegradable matter is biologically inert and passes through an activated sludge system in unchanged form. The inert soluble organic matter (S_I) leaves the system at the same concentration as it enters. Inert suspended organic matter in the wastewater influent (X_I) or produced via decay (X_p) becomes enmeshed in the activated sludge and is removed from the system via the sludge wastage.
- The biodegradable matter is divided into soluble readily biodegradable (S_S) and slowly biodegradable (X_s) substrate. Already here it should be stressed that some slowly biodegradable matter may actually be soluble. The readily biodegradable substrate is assumed to consist of relatively simple molecules that may be taken in directly by heterotrophic organisms and used for growth of new biomass. On the contrary, the slowly biodegradable substrate consists of relatively complex molecules that require enzymatic breakdown prior to utilization.
- Finally, heterotrophic biomass (X_{BH}) and autotrophy biomass (X_{BA}) are generated by growth on the readily biodegradable substrate (S_s) or by growth on ammonia nitrogen (S_{NH}). The biomass is lost via the decay process where it is converted to Xp and Xs (death regeneration). See Figure 1

The total COD balance of ASM1 is defined by

$$COD_{tot} = S_I + S_S + X_I + X_S + X_{BH} + X_{BA} + X_P \quad (2.1)$$

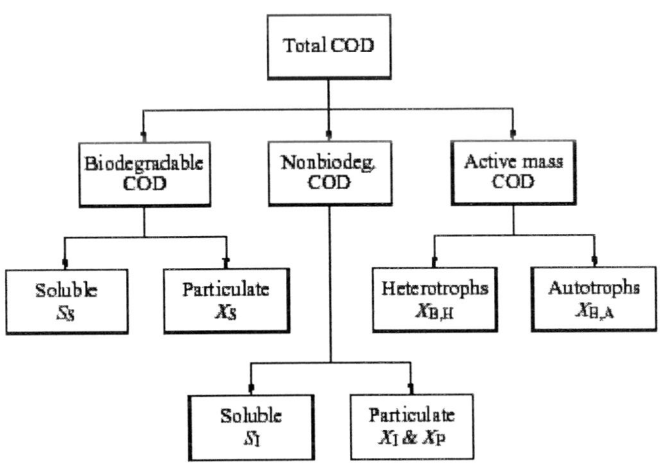

Figure 1. COD components in ASM1.

- **Nitrogen components in ASM1**

Similar to the organic matter, total nitrogen can be subdivided based on (1) solubility, (2) biodegradability and (3) biodegradation rate:

- Total nitrogen can be subdivided into soluble (S) and particulate (X) components.

- The nitrogen is divided into non-biodegradable matter and biodegradable matter. The non-biodegradable particulate organic nitrogen (X_{NI}) is associated with non-biodegradable particulate COD (X_I or X_P), whereas the soluble non-biodegradable organic nitrogen (S_{NI}) is assumed to be negligible and therefore not incorporated into the model.
- The biodegradable nitrogen is subdivided into ammonia nitrogen (S_{NH}), nitrate + nitrite nitrogen (S_{NO}), soluble organic nitrogen (S_{ND}) and particulate organic nitrogen (X_{ND}). The particulate organic nitrogen is hydrolysed to soluble organic nitrogen in parallel with hydrolysis of the slowly biodegradable organic matter (Xs) (either present in the wastewater or produced via the decay process). The soluble organic nitrogen is converted to ammonia nitrogen via ammonification. Ammonia nitrogen serves as the nitrogen source for biomass growth (the parameter i_{XB} indicates the amount of nitrogen incorporated per COD unit). Finally, the autotrophic conversion of ammonia results in nitrate nitrogen (S_{NO}), which is considered to be a single step process in ASM1. See Figure 2.

The total nitrogen balance for the components in ASM1 is defined by

$$N_{tot} = S_{NH} + S_{ND} + S_{NO} + X_{ND} + X_{NI} + i_{XB}*(X_{BH} + X_{BA}) + i_P*X_P \tag{2.2}$$

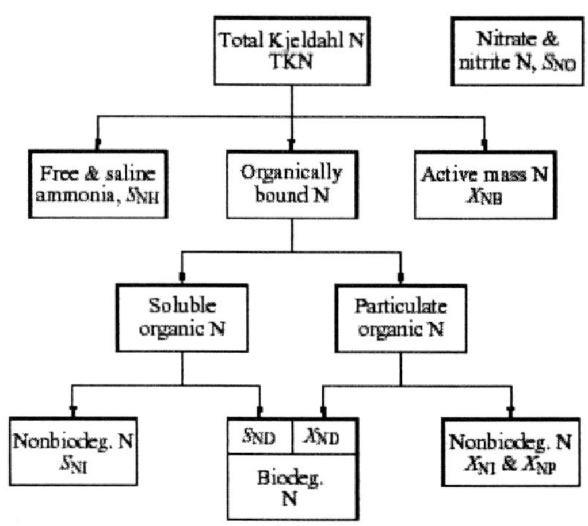

Figure 2. Nitrogen components in ASM1.

- **Processes in ASM1**

Basically there are four different main processes defined in ASM1:

- Growth of biomass
- Decay of biomass
- Ammonification of organic nitrogen
- Hydrolysis of particulate organic matter.

2.1.1. Aerobic Growth of Heterotrophic Biomass

Growth takes place by degradation of soluble readily biodegradable substrate (*Ss*) under the consumption of oxygen (*So*). Ammonia nitrogen (S_{NH}) is incorporated into cell mass. Both the concentrations of *Ss* and *So* may be rate limiting for the growth process. The Monod relationship is used to describe the growth of heterotrophic and autotrophic organisms.

- **Anoxic growth of heterotrophic biomass (denitrification)**

In the absence of oxygen the heterotrophic organisms are capable of using nitrate as the terminal electron acceptor with SS as substrate resulting in biomass growth and nitrogen gas. The same Monod kinetics as used for aerobic growth is applied except that the kinetic rate expression is multiplied by a correction factor η_g (<1). This factor is accounting for the fact that the anoxic substrate removal rate is slower compared to aerobic conditions. This can either be caused by a lower maximum growth rate or because only a fraction of the heterotrophic biomass is able to denitrify. Furthermore, anoxic growth is inhibited when oxygen is present which is described by the switching function $K_{OH}/(K_{OH}+S_O)$. The coefficient K_{OH} has the same value as in the expression for aerobic growth. Thus, as aerobic growth declines, the capacity for anoxic growth increases.

- **Aerobic growth of autotrophic biomass (nitrification).**

Ammonia nitrogen (S_{NH}) is oxidized to nitrate resulting in production of autotrophic biomass. Furthermore, a part of the S_{NH} is also incorporated in the autotrophic cell mass. As for heterotrophic growth the concentrations of S_{NH} and S_O can be rate limiting for the process. Nitrification has a considerable effect on the alkalinity (SALK).

- **Decay of heterotrophic biomass.**

The death regeneration concept of Dold (1980) [7] is applied to describe the different reactions that take place when organisms die. The traditional endogenous respiration concept describes how a fraction of the organism mass disappears to provide energy for maintenance. However, in the death regeneration concept oxygen is not directly associated with microbial decay. Decay is assumed to result in the release of slowly biodegradable substrate that is recycled back to soluble substrate and used for more cell growth. Thus, the oxygen utilization normally associated directly with decay is calculated as if it occurs indirectly form growth of new biomass on released substrate. A parallel conversion of organic nitrogen to ammonia nitrogen occurs. It should be noted that the magnitude of the decay coefficient used in this approach is different from that of the endogenous respiration. In endogenous respiration the loss of one unit of biomass COD leads to the utilization of one unit of oxygen minus the COD of the inert particulate products that are formed. However, in the death regeneration model the loss of one biomass COD unit results in the ultimate formation of one unit of COD due to the formed readily biodegradable substrate minus the formed inert particulate products. When the readily biodegradable COD is used for cell synthesis, only a fraction of a unit of oxygen (determined by the yield) will be required because of the energy incorporated into the cell mass. That cell mass undergoes in turn decay etc. before the unit of oxygen is finally removed.

- **Decay of autotrophic biomass.**

The decay of autotrophs is described similar to the heterotrophic decay process.

- **Ammonification of soluble organic nitrogen (SND).**

Biodegradable soluble organic nitrogen (SND) is converted to ammonia nitrogen (S_{NH}) in a first order process. Hydrogen ions consumed in this conversion process result in alkalinity change.

- **Hydrolysis**

Slowly biodegradable substrate (Xs) enmeshed in the sludge is broken down producing readily biodegradable substrate (Ss). The degradation of slowly biodegradable matter has appeared rather important to realistic modeling of activated sludge systems because it is primarily responsible for realistic electron acceptor profiles [7]. This process is modeled on the basis of surface reaction kinetics and occurs only under aerobic and anoxic conditions. The hydrolysis rate is reduced under anoxic conditions in the same way as anoxic growth, by applying a correction factor η_h (<1). The rate is also first order with respect to the heterotrophic biomass concentration present but saturates, as the amount of entrapped substrate becomes large in proportion to the biomass.

2.1.2. Restrictions of ASM1

A number of restrictions concerning ASM1 are summarized below [2]:

- The system must operate at constant temperature.
- The pH is constant and near neutrality. It is know that the pH has an influence on many of the parameters, however only limited knowledge is available to be able to express these possible influences. Consequently, a constant pH has been assumed. The inclusion of alkalinity in the model, however, does allow for detection of pH problems.
- No considerations have been given to changes in the nature of the organic matter within any given wastewater fractions (e.g. the readily biodegradable substrate). Therefore, the parameters in the rate expressions have been assumed to have constant values. This means that only concentration changes of the wastewater components can be handled whereas changes in the wastewater character can not.
- The effects of nutrient limitations (e.g. N and P) on the cell growth have not been considered. It is, however, easy to add limitation terms in the model if needed.
- The correction factors for denitrification are fixed and constant for a given wastewater, even though it is possible that their values are depending on the system configuration.

- The parameters for nitrification are assumed to be constant and to incorporate any inhibitory effects that wastewater constituents may have on them.
- The heterotrophic biomass is homogeneous and does not undergo changes in species diversity with time. This assumption is inherent to the assumption of kinetic parameters. This means that any changes in substrate concentration gradients, reactor configuration, etc. on sludge settleability are not considered.
- The entrapment of particulate organic matter in the biomass is assumed to be instantaneous.
- The hydrolysis of organic matter and organic nitrogen are coupled and occur simultaneously with equal rates.
- The type of electron acceptor present does not affect the loss of biomass by decay.
- The type of electron acceptor does not affect the heterotrophic yield coefficient.
- ASM1 is developed for simulation of treatment of municipal wastewater, and it is therefore not advised to apply the model to systems where industrial contributions dominate ht characteristics of the wastewater.
- ASM1 does not include processes that describe behaviors under anaerobic conditions. Simulations of systems with large fractions of anaerobic reactor volume may therefore lead to errors.
- ASM1 can not deal with elevated nitrite concentrations.
- ASM1 is not designed to deal with activated sludge systems with very high load or small sludge retention time (SRT) (<1 day).

2.2. Activated Sludge Model No. 3 (ASM3)

In the development of ASM3 some limitations of ASM1 were evaluated, and combined with the experiences gained with the application of ASM1 the following list of "defects" of ASM1 was defined by Gujer [5]:

- ASM1 does not include expressions to deal with nitrogen and alkalinity limitations.
- ASM1 considers biodegradable soluble and particulate organic nitrogen as model components. These can, however, not easily be measured and may in most cases unnecessarily complicate the use of ASM1.
- The ammonification kinetics can not be easily quantified, and moreover this process is typically rather fast and does therefore not affect model predictions significantly.
- ASM1 differentiates between inert suspended organic matter present in the influent wastewater and produced within the activated sludge process. In reality, however, it is impossible to distinguish between these two components.
- Hydrolysis has a rather dominating effect upon the predictions of the oxygen consumption and denitrification by heterotrophic organisms. In reality this process includes different coupled processes such as hydrolysis, lysis and storage of substrates. Therefore, the identification of the kinetic parameters of this combined process is difficult.

- Elevated concentrations of readily biodegradable organic substrates can lead to storage of poly-hydroxy-alkanoates, lipids or glycogen. This process is not included in ASM1.
- ASM1 does not include the possibility to differentiate between decay rates of nitrifiers under aerobic and anoxic conditions. This may lead to problems with the predictions of the maximum nitrification rates in cases of high SRT and high fractions of anoxic reactor volumes.

In ASM3, the conversion processes of both groups of organisms (autotrophs and heterotrophs) are clearly separated in ASM3 [6], whereas the decay regeneration cycles of the autotrophs and heterotrophs are strongly interrelated in ASM1. This change of decay concept (and introduction of the storage step) means that there exist more "entry" points for oxygen utilization resulting in, at some points, easier separation and characterization of the processes. Second, there is a shift of emphasis from hydrolysis to storage of organic matters. This gives a change in how wastewater characterization should be defined since the separation between *Ss* and *Xs* now should be based on the storage process rather than on the growth process. Still, the separation remains somewhat based on biodegradation rates. In ASM3 hydrolysis is obviously of a less dominating importance for the rates of oxygen consumption since only hydrolysis of *Xs* in the influent is considered.

- **COD components in ASM3**

The COD components in ASM3 are basically defined in the same way as in ASM1. Only the separation between inert suspended organic matter in the wastewater influent (X_I) and produced via the decay process (X_P) is no longer maintained, and, second, the component X_{STO} is introduced. The substrate Ss goes through the storage process but is basically still biodegradable. Thus, the total COD balance is defined by

$$COD_{tot} = S_I+S_S+X_I+X_S+X_H+X_A+X_{STO} \tag{5.1}$$

- **Nitrogen components in ASM3**

The nitrogen balance in ASM3 is simplified compared to ASM1, since the soluble and particulate organic nitrogen components are no longer considered. Furthermore, a nitrogen gas component (S_{N2}) is included allowing for a closed nitrogen mass balance. The nitrogen incorporated in S_I, S_S, X_I, X_S, and the biomass is defined in ASM3 as a fraction of these components. This fraction is consumed or produced when the corresponding COD fraction is formed or degraded respectively. The total nitrogen balance for the components in ASM3 is defined by

$$N_{tot} = S_{NH}+S_{NO}+S_{N2}+i_{NSI}*S_I+i_{NSS}*X_S+i_{NBM}*(X_H+X_A)+i_{NXI}*X_I \tag{5.2}$$

- **Processes in ASM3**

In ASM3, there are also four basic processes, however, slightly different from ASM1 [5]:

- Storage of readily biodegradable substrate
- Growth of biomass
- Decay of biomass
- Hydrolysis of particulate organic matter

- **Aerobic storage of readily biodegradable substrate**

This process describes the storage of readily biodegradable substrate (S_S) in the form of X_{STO} with the consumption of oxygen. As stated above, it is assumed that all SS first becomes stored material before use for cell growth. It is realized that this is not in accordance with reality. However, no model is currently available to predict the separation of SS into direct growth and storage. Gujer [5] therefore suggested to apply a low storage yield (Y_{STO}) and a higher growth yield (Y_H) to approximate direct growth.

- **Anoxic storage of readily biodegradable substrate**

This process is identical to the aerobic storage, only is nitrate used as terminal electron acceptor instead of oxygen. Furthermore, a correction factor (η_{No}) is applied to indicate that only a fraction of the heterotrophic biomass may be capable of denitrifying.

- **Aerobic growth of heterotrophs**

Aerobic heterotrophic growth takes place by degradation of X_{STO} with the consumption of oxygen (S_O). Ammonia nitrogen (S_{NH}) is incorporated into cell mass, as described for ASM1.

- **Anoxic growth of heterotrophs (denitrification)**

Anoxic growth is similar to aerobic growth but respiration is based on denitrification. Again, a correction factor (η_{No}) is applied to account for the observation of reduced anoxic respiration rates compared to aerobic respiration.

- **Aerobic growth of autotrophs (nitrification)**

This process is described similarly to ASM1.

- **Aerobic decay of heterotrophs**

The energy requirements not associated with growth but including maintenance, analysis, etc. are described by endogenous respiration in ASM3 according to simple first order reaction kinetics.

- **Anoxic decay of heterotrophs**

ASM3 allows for a description of anoxic decay in a similar way as the aerobic decay process.

- **Aerobic and anoxic decay of autotrophs**

The decay of autotrophs is described in the same way as the heterotrophic decay process.

- **Aerobic and anoxic respiration of storage products**

These processes are analogous to endogenous respiration and ensure that the storage product X_{STO} decays together with the biomass.

- **Hydrolysis**

Just as in ASM1 hydrolysis is responsible for the breakdown of slowly biodegradable substrate (X_S) to readily biodegradable substrate (S_S). However, in ASM3 hydrolysis is assumed to be electron donor independent, and as stressed above the hydrolysis does not play the same role as in ASM1.

2.2.1. Restrictions of ASM3

The restrictions are the same as for ASM1, except for the restriction stating that the type of electron acceptor does not affect the biomass decay.

3. Pattern Recognition

The subject may be said to have begun in about 1955 [8] when the banks began to be seriously interested in automatic sorting. The character recognition problem was greatly simplified by restricting attention to the numerals and a few other symbols. Recently interest in scientist applications has broadened out; in fact potentially useful approach for interpreting the data generated by chemical and biochemical processes. This approach is particularly useful when data are abundant, and expert knowledge is lacking. Pattern recognition may also complement the expert system approach by carrying out the low level assessment of process trends not easily captured in rules.

Wastewater processing by means of activated sludge is the most widespread sewage treatment process. The activated sludge process is with the further complication of operating in a widely changing environment where process conditions may be appreciably altered by unmeasurable inputs.

3.1. Establishing a Pattern Base of the Process

The most important task in knowledge-based recognition procedure is the conversion of input process variables into features. The second is the translation of the classification logic expressed initially in natural linguistic form into a transparent recognition algorithm that yields from the structural variables membership to different classes. For the initial representation of the recognition algorithm, we consider the distance, F, between two process state variables, which is determined by 2-norm of the difference between all normalized clustering centroids of patters, i.e:

$$F = \sum_{j=1}^{n} \left\| c_{rj} - c_{hj} \right\|^{2} \tag{3.1}$$

The exponent is the equation 3.1 is superscript not subscript

Where the centroids, c, is calculated using the traditional equation center of gravity. The amount of information obtained from a process is limited at the initial stage. However, with the passage of time, historical data will be more and more plentiful, and understanding of the process will be more and more clear [8]. The Table 2 shows the membership fuzzy values used in this analysis. In the Table 2, the Gaussian shaped function was used. The functional definition can readily be adapted to a change in the universe. The functional definition was expressed as

$$\mu_f = \exp\left[\frac{-(x - u_f)^2}{2\sigma_f^2} \right] \tag{3.2}$$

Where u_f is the Mean and σ_f is the Standard deviation

Table 2. Membership fuzzy values

Variable	Segments	u_f	σ_f	Primary fuzzy sets
S_I: soluble inert organic (g COD m^{-3})	[0.0, 5.0] [5.0, 22.5] [22.5, 35.0]	-0.5 0.05 0.5	0.3 0.3 0.2	LOW ZERO HIGH
S_S: Readily biodegradable substrate (g COD m^{-3})	[0.0, 40.0] [40.0, 84.0] [84.0, 130.0]	-0.1 0.3 0.57	0.3 0.3 0.2	LOW ZERO HIGH
S_{NH}: Ammonium (gN m^{-3})	[0.0, 2.2] [2.2, 11.1] [11.1, 20.0]	-0.2 0.07 0.55	0.4 0.3 0.3	LOW ZERO HIGH

The main difference between ASM1 and ASM3 is the recognition of the importance of storage polymers in the heterotrophic conversions in the activated sludge processes in ASM3. The aerobic storage process in ASM3 describes the storage of the readily biodegradable substrate (S_S) into a cell internal component (X_{STO}). The energy required for this process is obtained via aerobic respiration. This internal component is then subsequently used for growth. In ASM3 it is assumed that all Ss is first taken up and stored prior to growth. Thus, a division of the storage and growth process, allowing growth to take place on external substrate directly, is not considered. Furthermore, the death regeneration concept is replaced by endogenous respiration, which is closer to the phenomena observed in reality. Endogenous respiration can readily be obtained from a simple batch test. Also, ASM3 allows a differentiation between aerobic and anoxic decay.

The process states need to be transformed into an algorithm. It allows translation of the symbolic information into a numeric algorithm improving considerably their real response. For this, the following norm was adopted

$$\begin{aligned} \sum w_{ij}\mu_i &= dc_i \\ \sum w_{ij} &= CF_i \end{aligned} \tag{3.3}$$

Where w_{ij} are weight coefficients expressing the importance of the corresponding fuzzy membership, FC is Maximum value in the rule premise and *dci* are functions formalizing the membership set (credibility). The norm implies that every fact contributes additively and with different weight to the overall certainty of the rule construction. An important step in the creation of the recognition procedure consists of determining the unknown weight coefficients w_{ij}. They can be estimated by linear regression using a training set composed of a time-series of the structural variables and the credibility coefficient (dc_i). The maximum value of the credibility coefficient is 1. The proper regression procedure was constructed using [9]. The partition of the continuous universe requires a priori knowledge of the input/output space.

The Table 3 summarizes the kinetic parameters used.

Table 3. Kinetic parameters

Kinetic parameters	Values at 20C
K_S (gCOD m^{-3})	20.0
K_{OH} (gO_2 m^{-3})	0.21
K_{NO} (gNO m^{-3})	0.51
b_H (day^{-1})	0.62
η_g	0.8
η_h	0.4

The rest of the values used in this chapter can be found in [1]. The software of pattern of recognition was developed in MATLAB. The tests were carried out by simulation (off line) of historical data obtained from the plant. This part of the project consumed plenty time because the linear regression used in the training sections needed too much data that we obtained from the plant, thus that we included data generated from our own simulator. A comparative result indicates as the pattern of recognition has a good accuracy with respect to the mathematical models ASM1 and ASM3. The results are summarized in the Tables 4 and 5. Here we are shown one case where the knowledge-based approach can be useful in predicting the behavior of the settler. The use and validation of the Pattern recognition approach in the prediction of the typical characteristics of settled domestic sewage indicated that this method is good enough. However, in uncertain environments, 100% success can be never be reached such is the case of the prediction of the variables. We consider the continuous accumulation of the knowledge concerning the behavior of the system can improved the precision of the method as demonstrated the Tables 4 and 5.

Table 4. Comparative study of predicted and real values. ASM1

Variable	Predicted values using a data base with 1000 samples	Predicted values using a data base with 2000 samples	Predicted values using a data base with 3000 samples	Real Values
S_I: soluble inert organic (g COD m^{-3})	35	38	40	41
S_S: Readily biodegradable substrate (g COD m^{-3})	114	117	123	123
S_{NH}: Ammonium (gN m^{-3})	25	27	30	30

Table 5. Comparative study of predicted and real values. ASM3

Variable	Predicted values using a data base with 1000 samples	Predicted values using a data base with 2000 samples	Predicted values using a data base with 3000 samples	Real Values
S_I: soluble inert organic (g COD m^{-3})	36	37	41	41
S_S: Readily biodegradable substrate (g COD m^{-3})	115	119	122	123
S_{NH}: Ammonium (gN m^{-3})	26	29	31	30

References

[1] Ardern, E. and Lockett, W.T. "Experiments on the oxidation of sewage without the aid of filters", *Journal of the Society of Chemical Industries*, **33**, 523, 1914.

[2] Henze, M., Grady Jr, C. P. L., Gujer, W., Marais, G. v. R., Matsuo, T. (1986). *Activated sludge model no 1.* International Association on Water Pollution Research and Control. *Scientific and Technical Reports* No **1**. ISSN: 1010-707X.

[3] Henze, M., Gujer, W., Mino, T., Matsuo, T., Wentzel, M. C., Marais, G. v. R., Van Loosdrecht, M. C. M. (1995). *Activated sludge model no* **2**. *Scientific and Technical Reports* No **3**. IAWQ.

[4] Forster, C. F. "Wastewater Treatment and Technology", 1st Edition, Thomas Telford Publishing, Thomas Telford Ltd., London, 2003, 25-27 & 204.

[5] Gujer, W., Henze, M. (1991), "Activated Sludge Modeling and Simulation". *Wat. Sci. Tech.*, vol. 23, pp. 1011-1023.

[6] Ekama, G.A., P.L. Dold and G.V.R Marais (1986) Procedures for determining COD fractions and the maximum specific growth rate of heterotrophs in activated sludge systems, *Wat. Sci. Tech.* **18** (6), pp. 91-114.

[7] Dold, P.L. and G.v. Marais (1986) Evaluation of the general activated sludge model proposed by the IAWPRC task group, *Wat. Sci. Tech.* **18** (6) pp. 63-89.

[8] Negoita, C. V. and Ralescu D., On Fuzzy Optimization. *Kybernetes* **6** (1977) 139-96.

[9] Sengupta, A., Pal, T. K and Chakraborty, D. Interpretation Of Inequality Constraints Involving Interval Coefficients and A Solution To Interval Linear Programming. *Fuzzy Sets and Systems* **119** (2001) 129-38

In: Pattern Recognition in Nanoscience...
Editor: Diane M. Perez, pp. 59-96
ISBN: 1-60021-715-X

Chapter 3

ESTIMATION OF MOTION OR VIBRATION FROM MOTION BLURRED IMAGES

Shigan Wang, Baiqing Guan, Qian Li and Lin He
Shanghai Jiao Tong University, Shanghai 200240, P.R.China

Abstract

Traditional vision-based methods usually estimate motion or vibration from large sequence of unblurred images by employing optical flow algorithms on the image data. However, the image sequence tends to get blurred when the displacement of the object is large within the exposure time of the camera. In this chapter, we introduce a novel kind of approaches based on motion-blurred-images for measuring motion or vibration parameters. This group of approaches utilizes the motion blur information in the motion blurred images to measure motion or vibration parameters. This chapter derives two classes of models to represent the motion blur information from the motion-blurred images and measure the motion or vibration parameters from the motion blur: the first one is based on image features in the spatial domain and the second one is based on that in the frequency domain.

In spatial domain, we employ geometric moments to extract the motion blur cue for measuring motion or vibration parameters. From our previous work, we derive a set of nonlinear equations of the motion parameters from the relationship between the geometric moments of translational motion blurred images and the displacement function of the motion. Thus, we calculate the parameters of the motion by solving these nonlinear equations.

In frequency domain, the spectrum of translational motion blurred image is multiplied by a Sinc function for translation or a zero-order Bessel function for high-frequency vibration, which has zeros at positions related to the length of the boxcar. When these zeros are found by image processing method, the parameters of image motions can be estimated.

For rotational motion, we represent the images in polar coordinate system. Obviously, estimation of the angular velocity of a monaxial rotation in polar coordinates is equivalent to estimating the velocity of 1D tranlstional motion in Cartesian coordinate system.

These techniques for measuring motion or vibration from motion blur can measure uniform motion, accelerated motion, vibration from only one motion blurred image and an unblurred image for reference or from only two successive frames of motion blurred images, which is almost impossible for high-speed video or stroboscopic photography. We employ examples with both synthetic and real motion blurred images to illustrate the performance of the proposed methods in estimating motion.

1. Introduction

Motion or vibration analysis is very important in many applications, such as motion control, fault diagnosis, dynamic testing of mechanical system, intelligent robot, intelligent monitoring, computer aided diagnosis system, etc. Conventional motion or vibration measurements are piezoelectric velocimeters or accelerometers. However, piezoelectric sensors are based on mechanical contact, which may perturb the operation of objects or destructing the objects. Over the past decades, non-contact approaches are developed to provide non-destruct measurement of motions or vibrations. Existing non-contact measurements are usually based optical techniques, such as electronic speckle pattern interferometry (Osten, et al., 1999), holographic techniques (Salbut, 2001; Borza, 2004), laser vibrometry (Turner, Hatwell and MacDonald, 1999; Link and von Martens, 1998; Rothberg and Bell, 2004), etc. Among all non-contact measurements of motion or vibration, a flexible and low-cost one is vision-based approach, which uses a CCD camera to acquire images or image sequences of the moving object and analyze the motion or vibration by applying image processing algorithms to the image data. Previous vision-based techniques for vibration measurement can be mainly categorized into two groups:

(1) Sequential-unblurred-image-based methods. Sequential-unblurred-image-based methods estimate the motion or vibration parameters by optical flow or other motion estimation algorithms (Barron et al., 1994; Ye et al., 2003; Arredondo et al., 2004; Ming et al., 2002) on a large sequence of unblurred images with the assumption that the exposure time is infinitesimal and the motion within the exposure time can be neglect. Gebert et al. (2003) proposed an algorithm for the determination of angular rates and range from the spatio-temporal image gradients by employing optical flow constraints. Tigges et al. (1999) proposed a method to demonstrate vocal fold vibration by digital multi-plane kymography combined with digital high-speed recordings. A method for monitoring and real-time measuring the displacement of chosen points at bridge structures by a CCD camera with telephoto lens was introduced in (Olaszek, 1999). However, the exposure time of the camera is practically nonintinitesimal, and the motion or vibration of the object usually causes motion blur on the image plane, which invalidates the unblurred-image-based techniques. Further, for reciprocative motions, i.e. vibrations, the frame rate of the camera should be greater than Nyquist frequency to restore the vibrational signal. Therefore, to measure a motion or vibration accurately, unblurred-image-based methods should using high-speed cameras to acquire sufficient frames of unblurred images. When the speed of the motion or the vibrational frequency is very large, even the cameras with the highest frame rate are hardly competent. In this case, Freeman and his followers (Freeman, 1998; Hemmers, Mermelstein, Freeman, 1999; Hafiane, Petitgrand, Gigan et al., 2003) employ stroboscopic illumination to "freeze" the vibration object in images and acquire large sequences of unblurred images for measuring the vibrations of MEMS. In Freeman's approach, the quality and quantity of images acquired by the camera is not determined by the shutter speed and frame rate of the camera, but by the light pulse width and the minimum phase shift of light pulses.

(2) Motion-blurred-image-based approaches. While sequential-unblurred-image-based methods lay emphasis on acquiring large sequences of unblurred images by high-speed cameras or stroboscopic photography to realize motion or vibration analysis, motion-blurred-

image-based techniques analyze the motions from motion blurred images recorded by a general camera with noninfinitesimal exposure time. Presently, most investigations on motion blurred images focus on blurred image restoration (Stern and Kopeika, 1997; Stern, Kempner, Shukrun, et al., 2000;Yitzhaky, Boshusha, Levy, et al., 2000; Hadar, Adar and Cotter, et al., 1997; Ben-Ezra and Nayar, 2004). Several other authors (Flusser and Suk, 1998) presented blur invariants in both spatial and frequency domain, which is irrelative to image blur, to represent objects. These blur invariants can be used to estimate motion from blurred images by image registration (Zhang et al., 2000). Both image restoration techniques and blur-invariants-based approaches consider motion blur as degradation of images.

In this chapter, we introduce a novel kind of approaches based on motion blurred images for measuring motion or vibration parameters in industrial applications. These approaches regard motion blur as an important cue for motion or vibration estimation, and analyze motions from the motion blur information in motion blurred images directly. The presented approaches can be classified into two categories: the spatial-model-based method and the frequency-model-based one. In spatial domain, we employ geometric moments to extract the motion blur cue for measuring motion or vibration parameters. In frequency domain, we use the stripes formed by the zeros of the spectrum of the motion blurred image to estimate motion or vibration. The organization of the rest of this chapter is as follows:

In section 2, we introduce the mathematical models for motion blurred images. Section 3 describes the theory and application of the novel approaches for estimating motion or vibration parameters. Experimental results with both simulated and real motions or vibrations are also provided to validate the proposed approaches in this section. In section 4, a conclusion is given.

2. Mathematical Models of Motion Blurred Images

In this section, we will introduce the mathematical model of motion blurred images, which will be used in the presented algorithms for motion or vibration measurement.

2.1. General Model

According to the imaging mechanisms of the camera, the gray level of any image point is proportional to the integral of the light intensity received by the image point within the exposure time if there exists no gray level saturation. Suppose we have caught an unblurred image $f(x,y)$ with the exposure time $t_e^{(f)}$ when the object is static. The unblurred image $f(x,y)$ can be given by:

$$f(x,y) = K_f \int_0^{t_e^{(f)}} \eta_s(x,y)dt \qquad (1)$$

where $\eta_s(x,y)$ is the light intensity received by the image point (x,y) when the object is static, K_f is the proportional coefficient.

Similarly, let $\eta_m(x,y,t)$ be the light intensity received by the image point (x,y) at time t when the object is moving, we get the i-th motion blurred image $g^{(i)}(x,y)$:

$$g^{(i)}(x,y) = K_{g^{(i)}} \int_{t_s^{(i)}}^{t_e^{(g^{(i)})}} \eta_m(x,y,t)dt \tag{2}$$

where $K_{g^{(i)}}$ is the proportional coefficient when catching $g^{(i)}(x,y)$, $t_s^{(i)}$ denotes the time when the camera's shutter opening at the i-th sampling period, and $t_e^{(g^{(i)})}$ denotes the exposure time for the imaging system to acquire the motion blurred image $g^{(i)}(x,y)$.

Based on optical and kinematic principles, $\eta_m(x,y,t)$ and $\eta_s(x,y)$ satisfy:

$$\eta_m(x,y,t) = Trans(\eta_s(x,y),t) \tag{3}$$

where $Trans(\bullet,t)$ is an operator which is determined by the optical and kinematic constraints.

According to Eq.(1)-(3), the i-th motion blurred image $g^{(i)}(x,y)$ can be modeled by:

$$g^{(i)}(x,y) = \frac{K_{g^{(i)}}}{K_f t_e^{(f)}} \int_{t_s^{(i)}}^{t_e^{(g^{(i)})}} Trans(f(x,y),t)dt \tag{4}$$

2.2. Models of Images Blurred by Various Motions

In section 2.1, we introduce the general model of motion blurred images. In this chapter, we mainly deal with motion or vibration analysis in industrial applications. Therefore, we present some specific models for blurred images with various motions to estimate the parameters of motions or vibrations in different situations.

2.2.1. Basic Assumption

The general model shown in Eq. (4) can be very complicated if we have little knowledge about the object and environment. To simplify the models of motion blurred images, we make the following assumptions:

(1) Suppose the illumination is constant.

(2) The object surfaces are Lambertian surfaces.
With the above 2 assumptions, the light intensity reflected by the same point on the object surface keeps constant whether the object is moving or is static.

(3) The background is homogenously illuminated. Thus, the background can be considered as moving synchronously with the object.

Note that these basic assumptions can be satisfied in most industrial applications, we can build the following models of blurred images with various motions.

2.2.2. Model of Images Blurred by In-Plane Translational Motion

Many industrial applications involve only pure translational motions (Gregorio and Parenti-Castelli, 2002) or translational vibrations. Therefore, translational motion or vibration analysis is of great importance. Translational motion or vibration can be classified into two categories according to the position of the optical axis relative to the motion direction: (1) In-plane motion. For in-plane motion, the optical axis of the imaging system is perpendicular to the direction of the motion, i.e., the image plane is parallel to the motion direction. (2) Out-of-plane motion. The optical axis of the camera is parallel to the direction of the out-of-plane motion.

In this section, we give a mathematical model in spatial domain to describe the images blurred by in-plane translational motions.

For in-plane motion, the light intensity $\eta_m(x,y,t)$ received by the image point (x,y) at time t when the object is moving and the received light intensity $\eta_s(x,y)$ satisfy:

$$\eta_m(x,y,t)=\eta_s\left(x-d_x(t),y-d_y(t)\right) \tag{5}$$

where $d_x(t)$ and $d_y(t)$ are the displacements of the position of the object in the motion blurred image with regard to its initial position in the unblurred image $f(x,y)$ at time t in x and y directions, respectively.

From Eq. (5) and Eq. (4), we get the i-th image $g^{(i)}(x,y)$ blurred by in-plane translational motion:

$$g^{(i)}(x,y)=\frac{K_{g^{(i)}}}{K_f t_e^{(f)}}\int_{t_s^{(i)}}^{t_e^{\left(g^{(i)}\right)}} f\left(x-d_x(t),y-d_y(t)\right)dt \tag{6}$$

2.2.3. Model of Images Blurred by Monaxial Rotation

Suppose the rotational axis is parallel to the optical axis of the imaging system, then a monaxial rotation in polar coordinates is equivalent to an in-plane translational motion. Therefore, we can easily map the Cartesian model for in-plane tranlational motion blurred images to a spatial model of the image blurred by monaxial rotation in polar coordinates.

We introduce a polar coordinates (shown in Fig.1) to represent the image plane, the polar coordinates of the image point is (r,θ). The polar radius r and the polar angle of a Cartesian point (x,y) are given respectively by:

$$r=\sqrt{(x-x_c)^2+(y-y_c)^2} \quad (7)$$

$$\theta=\arctan\left(-(y-y_c)/(x-x_c)\right)\pm 2k\pi,\ k=0,1,2,\cdots. \quad (8),$$

where (x_c, y_c) are the Cartesian coordinates of the image point and the rotating center of the image.

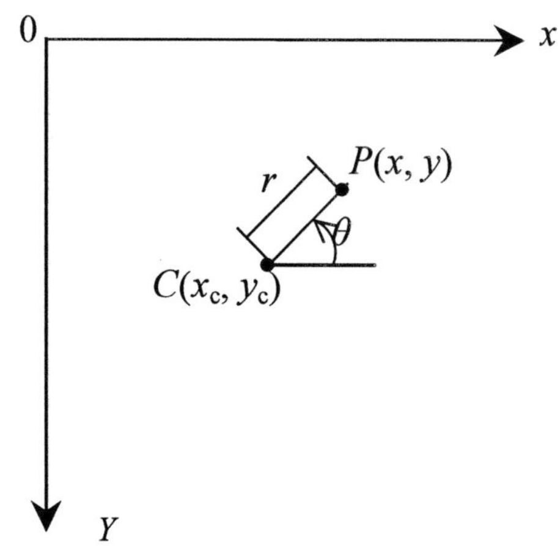

Figure 1. Polar coordinate system.

Let $f(r,\theta)$ be an unblurred polar image acquired when the object is static. The motion blurred polar image $g^{(i)}(r,\theta)$ can be modeled by.

$$g^{(i)}(r,\theta)=\frac{K_{g^{(i)}}}{K_f t_e^{(f)}}\int_{t_s^{(i)}}^{t_s^{(i)}+t_e^{\left(g^{(i)}\right)}} f\left(r,\theta-\Delta\theta(t)\right)\mathrm{d}t \quad (9)$$

where $\Delta\theta(t)$ is the angular displacement function of the object with regard to its initial position θ in the unblurred image $f(r,\theta)$.

Obviously, the images $f(r,\theta)$ and $g^{(i)}(r,\theta)$ in this polar coordinate system are periodic with respect to θ, and the period is 2π. Thus, $f(r,\theta)=f(r,\theta-2\pi)$, $g^{(i)}(r,\theta)=g^{(i)}(r,\theta-2\pi)$.

3. Estimating Motion or Vibration Parameters from Motion Blur

Psychophysical and biological investigations indicate that the human vision system and some other biological vision systems may have mechanisms to perceive motion from image blur (Burr, Ross and Morrone, 1986; Anderson and Van Essen, 1987). Therefore, a few authors consider that motion blur involves important information, which can be used in motion analysis or vibration measurement. Chen et al. (1996) described a special data acquisition scheme as well as a "motion-from-blur" algorithm. Burns and Helbig (1999) measured vibration amplitudes by analyzing the blurred zone extension at the periphery of the vibrating device in the time-averaged images acquired by optical microscopy. In this chapter, we introduce a group of novel approaches based on motion-blurred-images for measuring motion or vibration parameters. The presented approaches can be classified into two categories: the spatial-model-based method described in Section 3.1 and the frequency-model-based one introduced in section 3.2.

3.1. Estimating Motion or Vibration Parameters from Motion Blur by Using Spatial Models

Psychophysical and biological investigations show that there exist both temporal integration and spatial integration in biological vision systems (Warrant, 1999). Hence, we hypothesize that the process of calculating motion from blur may involve some spatial summation. In this section, we employ a generally used kind of spatial summation, i.e. geometric moment, to extract the motion blur information from the motion blurred images and estimate motion parameters from the motion blur cue. We at first analyze the relationship between the in-plane translational motion blur information and the geometric moments of the spatial models for in-plane tranlational motion blurred images and realize in-plane translational motions by using this relationship in section 3.1.1. Then, we map the approaches for analyzing in-plane translational motion to monaxial rotation measurements in section 3.1.2.

3.1.1. Estimating In-plane Translational Motions or Vibrations by Using Geometric Moments

As is mentioned in section 2.2.2, translational motions or vibrations exist in many industrial applications. In this section, we deal with in-plane translational motions. We at first introduce the definitions of the geometric moments of the motion blurred images and the moments of the displacement function of the motion, and then give a series of theorems to reveal the relationship between the moments of the motion blurred images and the motions. Then we describe the theory and applications for estimating in-plane tranlational motions or vibrations by using geometric moments. At last, we provide experimental results to validate the presented theory.

3.1.1.1. Geometric moments of the motion blurred images and the moments of the motions

The 2D geometric moment $M_{pq}^{(f)}$ of order $(p+q)$ of a gray-level image $f(x,y)$ is given by the integral:

$$M_{pq}^{(f)} = \int_{-\infty}^{\infty}\int_{-\infty}^{\infty} x^p y^q f(x,y)dxdy. \tag{10}$$

The 2D central moment $M_{pq}^{(f)}$ of order $(p+q)$ of a gray-level image $f(x,y)$ is given by:

$$\mu_{pq}^{(f)} = \int_{-\infty}^{\infty}\int_{-\infty}^{\infty}\left(x-\overline{x^{(f)}}\right)^p\left(y-\overline{y^{(f)}}\right)^q f(x,y)dxdy \tag{11}$$

where $\overline{x^{(f)}} = M_{10}^{(f)}/M_{00}^{(f)}$, $\overline{y^{(f)}} = M_{01}^{(f)}/M_{00}^{(f)}$.

From the definition of geometric moment in Eq. (10) and the translational motion blurred image function in Eq. (6), we get the $(p+q)$-th order moment $M_{pq}^{(g^{(i)})}$ of the translational motion blurred image $g^{(i)}(x,y)$:

$$M_{pq}^{\left(g^{(i)}\right)} = \int_{-\infty}^{\infty}\int_{-\infty}^{\infty} x^p y^q \left[\frac{K_{g^{(i)}}}{K_f t_e^{(f)}}\int_{t_s^{(i)}}^{t_e^{\left(g^{(i)}\right)}} f\left(x-d_x(t), y-d_y(t)\right)dt\right]dxdy. \tag{12}$$

The $(p+q)$-th order central moment $\mu_{pq}^{(g^{(i)})}$ of the translational motion blurred image $g^{(i)}(x,y)$ is given by:

$$\mu_{pq}^{\left(g^{(i)}\right)} = \int_{-\infty}^{\infty}\int_{-\infty}^{\infty}\left(x-\overline{x^{\left(g^{(i)}\right)}}\right)^p\left(y-\overline{y^{\left(g^{(i)}\right)}}\right)^q\left[\frac{K_{g^{(i)}}}{K_f t_e^{(f)}}\int_{t_s^{(i)}}^{t_e^{\left(g^{(i)}\right)}} f\left(x-d_x(t), y-d_y(t)\right)dt\right]dxdy \tag{13}$$

Furthermore, we define the $(p+q)$-th order moment $m_{pq}^{\left(d^{(i)}\right)}(\mathbf{d}_0)$ of the 2D translational motion $\mathbf{d}(t)$ with respect to $\mathbf{d}_0 = \left(d_{x0}, d_{y0}\right)^T$ during the time interval $\left[t_s^{(i)}, t_s^{(i)} + t_e^{\left(g^{(i)}\right)}\right]$ as:

$$m_{pq}^{\left(d^{(i)}\right)}(\mathbf{d}_0)=\frac{1}{t_e^{g^{(i)}}}\int_{t_s^{(i)}}^{t_s^{(i)}+t_e^{g^{(i)}}}\left(d_x(t)-d_{x0}\right)^p\left(d_y(t)-d_{y0}\right)^q dt. \tag{14}$$

The $(p+q)$-th order central moment $\mu_{pq}^{\left(d^{(i)}\right)}$ of the 2D translational motion $\mathbf{d}(t)$ during the time interval $\left[t_s^{(i)},t_s^{(i)}+t_e^{\left(g^{(i)}\right)}\right]$ is defined as:

$$\mu_{pq}^{\left(d^{(i)}\right)}=\frac{1}{t_e^{\left(g^{(i)}\right)}}\int_{t_s^{(i)}}^{t_s^{(i)}+t_e^{\left(g^{(i)}\right)}}\left(d_x(t)-\overline{d_x^{(i)}}\right)^p\left(d_y(t)-\overline{d_y^{(i)}}\right)^q dt \tag{15}$$

where $\overline{d_x^{(i)}}=\frac{1}{t_e^{\left(g^{(i)}\right)}}\int_{t_s^{(i)}}^{t_s^{(i)}+t_e^{\left(g^{(i)}\right)}} d_x(t)dt\,;\ \overline{d_y^{(i)}}=\frac{1}{t_e^{\left(g^{(i)}\right)}}\int_{t_s^{(i)}}^{t_s^{(i)}+t_e^{\left(g^{(i)}\right)}} d_y(t)dt.$

The geometric moments and central moments of the blurred image $g^{(i)}(x,y)$, the unblurred image $f(x,y)$ and the 2D translational motion $\mathbf{d}(t)$ are proved to satisfy Theorem1, Theorem 2 (Guan, 2005) and Theorem3.

Theorem 1: Let $M_{pq}^{\left(g^{(i)}\right)}$, $M_{pq}^{(f)}$ be the $(p+q)$-th order moments of the motion blurred image $g^{(i)}(x,y)$ and the unblurred image $f(x,y)$ respectively. Let $m_{pq}^{\left(d^{(i)}\right)}(\mathbf{0})$ be the $(p+q)$-th order moment of the 2D translational motion $\mathbf{d}(t)$ with respect to $(0,0)^T$. Then

$$M_{pq}^{\left(g^{(i)}\right)}=\frac{K_{g^{(i)}}t_e^{\left(g^{(i)}\right)}}{K_f t_e^{(f)}}\sum_{k=0}^{p}\sum_{l=0}^{q}\binom{p}{k}\binom{q}{l}m_{kl}^{\left(d^{(i)}\right)}(\mathbf{0})M_{p-k,q-l}^{(f)} \tag{16}$$

for any p and q.

Theorem 2: Let $m_{pq}^{\left(d^{(i)}\right)}(\mathbf{d}_0)$ be the $(p+q)$-th order moment of the 2D translational motion $\mathbf{d}(t)$ with respect to $\mathbf{d}_0=(d_{x0},d_{y0})$. Let $f'(x,y)=f(x-d_{x0},y-d_{y0})$. Then

$$M_{pq}^{\left(g^{(i)}\right)}=\frac{K_{g^{(i)}}t_e^{\left(g^{(i)}\right)}}{K_f t_e^{(f)}}\sum_{k=0}^{p}\sum_{l=0}^{q}\binom{p}{k}\binom{q}{l}m_{kl}^{\left(d^{(i)}\right)}(\mathbf{d}_0)M_{p-k,q-l}^{(f')}. \tag{17}$$

Theorem 3: Let $\mu_{pq}^{\left(g^{(i)}\right)}$, $\mu_{pq}^{(f)}$ be the $(p + q)$-th order moments of the motion blurred image $g^{(i)}\left(x, y\right)$ and the unblurred image $f(x, y)$ respectively. Let $\mu_{pq}^{\left(d^{(i)}\right)}$ be the $(p + q)$-th order central moment of the 2D translational motion $\mathbf{d}(t)$. Then

$$\mu_{pq}^{\left(g^{(i)}\right)} = \frac{K_{g^{(i)}} t_e^{\left(g^{(i)}\right)}}{K_f t_e^{(f)}} \sum_{k=0}^{p} \sum_{l=0}^{q} \binom{p}{k} \binom{q}{l} \mu_{kl}^{\left(d^{(i)}\right)} \mu_{p-k,q-l}^{(f)} \tag{18}$$

for any p and q.

Proof:

$$\mu_{pq}^{\left(g^{(i)}\right)} = \int_{-\infty}^{\infty}\int_{-\infty}^{\infty} \left(x - \overline{x^{\left(g^{(i)}\right)}}\right)^p \left(y - \overline{y^{\left(g^{(i)}\right)}}\right)^q \left[\frac{K_{g^{(i)}}}{K_f t_e^{(f)}} \int_{t_s^{(i)}}^{t_s^{(i)}+t_e^{\left(g^{(i)}\right)}} f\left(x - d_x(t), y - d_y(t)\right) dt\right] dxdy$$

According to Theorem 1, we get

$$\overline{x^{\left(g^{(i)}\right)}} = M_{10}^{\left(g^{(i)}\right)} \Big/ M_{00}^{\left(g^{(i)}\right)} = \left(\frac{K_{g^{(i)}} t_e^{\left(g^{(i)}\right)}}{K_f t_e^{(f)}} \left(M_{10}^{(f)} m_{00}^{\left(d^{(i)}\right)} + m_{10}^{\left(d^{(i)}\right)} M_{00}^{(f)}\right)\right) \Bigg/ \left(\frac{K_{g^{(i)}} t_e^{\left(g^{(i)}\right)}}{K_f t_e^{(f)}} M_{00}^{(f)}\right)$$

$$= M_{10}^{(f)} \Big/ M_{00}^{(f)} + m_{10}^{\left(d^{(i)}\right)} \Big/ m_{00}^{\left(d^{(i)}\right)} = \overline{x^{(f)}} + \overline{d_x^{(i)}}$$

$$\overline{y^{\left(g^{(i)}\right)}} = M_{01}^{\left(g^{(i)}\right)} \Big/ M_{00}^{\left(g^{(i)}\right)} = \overline{y^{(f)}} + \overline{d_y^{(i)}}$$

Therefore,

$$\mu_{pq}^{\left(g^{(i)}\right)} = \frac{K_{g^{(i)}}}{K_f t_e^{(f)}} \int_{t_s^{(i)}}^{t_s^{(i)}+t_e^{\left(g^{(i)}\right)}} \int_{-\infty}^{\infty}\int_{-\infty}^{\infty} \left(x + d_x(t) - \overline{x^{\left(g^{(i)}\right)}}\right)^p \left(y + d_y(t) - \overline{y^{\left(g^{(i)}\right)}}\right)^q f(x,y) dxdydt$$

$$= \frac{K_{g^{(i)}}}{K_f t_e^{(f)}} \int_{t_s^{(i)}}^{t_s^{(i)}+t_e^{\left(g^{(i)}\right)}} \int_{-\infty}^{\infty}\int_{-\infty}^{\infty} \left(x + d_x(t) - \overline{x^{(f)}} - \overline{d_x^{(i)}}\right)^p \left(y + d_y(t) - \overline{y^{(f)}} - \overline{d_y^{(i)}}\right)^q f(x,y) dxdydt$$

$$= \sum_{k=0}^{p} \sum_{l=0}^{q} \binom{p}{k} \binom{q}{l} \int_{-\infty}^{\infty}\int_{-\infty}^{\infty} \left(x - \overline{x^{(f)}}\right)^k \left(y - \overline{y^{(f)}}\right)^l f(x,y) dxdy$$

$$\times \frac{K_{g^{(i)}}}{K_f t_e^{(f)}} \int_{t_s^{(i)}}^{t_s^{(i)}+t_e^{\left(g^{(i)}\right)}} \left(d_x(t) - \overline{d_x^{(i)}}\right)^{p-k} \left(d_y(t) - \overline{d_y^{(i)}}\right)^{q-l} dt$$

$$= \frac{K^{\left(g^{(i)}\right)} t_e^{\left(g^{(i)}\right)}}{K_f t_e^{(f)}} \sum_{k=0}^{p} \sum_{l=0}^{q} \binom{p}{k} \binom{q}{l} \mu_{kl}^{(f)} \mu_{p-k,q-l}^{\left(d^{(i)}\right)}$$

3.1.1.2. Fundamental Theory for Estimating in-Plane Translational Motions by Using Geometric Moments

Theorem 1, 2 and 3 (shown in Eqs. (13-15)) reveal the relationship between the moments of the motion blurred images and the moments of the translational motion. According to these theorems, we can extract the motion blur information from the motion blurred images, and then estimate the motion parameters from motion blur by using a motion blurred image and an unblurred image for reference or using two successive frames of motion blurred images.

(1) Translational motion estimation from a motion blurred image and an unblurred image by using geometric moments

According Theorem 1, we can calculate the moments $m_{pq}^{\left(d^{(i)}\right)}(\mathbf{0})$ of the translational motion iteratively by using the geometric moments of the motion blurred image and an unblurred reference image, as is shown in Eq. (19):

$$m_{pq}^{\left(d^{(i)}\right)} = \left[\frac{K_f t_e^{(f)}}{K_{g^{(i)}} t_e^{\left(g^{(i)}\right)}} M_{pq}^{\left(g^{(i)}\right)} - \sum_{k=0}^{p-1}\sum_{l=0}^{q-1} \binom{p}{k}\binom{q}{l} m_{kl}^{\left(d^{(i)}\right)} M_{p-k,q-l}^{(f)} - \sum_{k=0}^{p-1} \binom{p}{k} m_{kq}^{\left(d^{(i)}\right)} M_{p-k,0}^{(f)} - \sum_{l=0}^{q-1} \binom{q}{l} m_{pl,}^{\left(d^{(i)}\right)} M_{0,q-l}^{(f)} \right] / M_{00}^{(f)} \tag{19}$$

where $m_{pq}^{\left(d^{(i)}\right)}$ denotes $m_{pq}^{\left(d^{(i)}\right)}(\mathbf{0})$.

Further, suppose the displacement function $\mathbf{d}(t)$ of the translational image motion takes the following takes the parametric form:

$$\mathbf{d}(t) = \mathbf{d}_0 + \Gamma\left(\mathbf{Para}_{2N}, \left(t - t_s^{(i)}\right)\right), t \in \left(t_s^{(i)}, t_s^{(i)} + t_e^{\left(g^{(i)}\right)}\right) \tag{20}$$

where $\mathbf{Para}_{2N} = \begin{bmatrix} para_{x1} & \cdots & para_{xk} & \cdots & para_{xN} \\ para_{y1} & \cdots & para_{yk} & \cdots & para_{yN} \end{bmatrix}$ is the $2 \times N$ parametric matrix of the displacement $\mathbf{d}(t)$ of the motion, $\mathbf{Para}_{2N}$ keeps invariant within the time interval $\left[t_s^{(i)}, t_s^{(i)} + t_e^{\left(g^{(i)}\right)}\right]$; Γ is a function of $\mathbf{Para}_{2N}$ and t.

Thus, the $(p+q)$-th moment $m_{pq}^{\left(d^{(i)}\right)}$ of the motion can be expressed by:

$$m_{pq}^{\left(d^{(i)}\right)} = \Phi_{pq}\left(\mathbf{Para}, \mathbf{d}_0\right) \tag{21}$$

where Φ_{pq} is a nonlinear function of $\mathbf{Para}$ and $\mathbf{d}_0$. Equations (21) for different p or q are independent of each other. Thus, the motion parameters in the matrix $\mathbf{Para}$ can be estimated by solving the following nonlinear equations:

$$\Phi_{pq}\left(\mathbf{Para}, \mathbf{d}_0\right) - m_{pq}^{\left(d^{(i)}\right)} = 0, \left(0 \le p \le m, 0 \le q \le n\right) \tag{22}$$

where $m_{pq}^{\left(d^{(i)}\right)}$ $(0 \le p \le m, 0 \le q \le n)$ can be computed by Eq. (19) if a motion blurred image and an unblurred reference image are given, where $m \ge N_x + 1$ and $N \ge N_y + 1$ (N_x and N_y are the numbers of nonzero items in $\mathbf{Para}_x$ and $\mathbf{Para}_y$).

In this chapter, we solve the nonlinear equations shown in Eqs. (22) by searching for the minimum of the squared error between $\Phi_{pq}(\mathbf{Para}, \mathbf{d}_0)$ and $m_{pq}^{\left(d^{(i)}\right)}$ for any p and q, as is shown in Eq. (23):

$$\begin{aligned} \left\{\hat{\mathbf{Para}}, \hat{\mathbf{d}}_0\right\} &= \arg\min_{\mathbf{Para}, \mathbf{d}_0} \varepsilon\left(\mathbf{Para}, \mathbf{d}_0\right) \\ &= \arg\min_{\mathbf{Para}, \mathbf{d}_0} \left(\sum_{p=0}^{m} \sum_{q=0}^{n} \left\| \Phi_{pq}\left(\mathbf{Para}\right) - m_{pq}^{\left(d^{(i)}\right)} \right\|^2 \right) \end{aligned} \tag{23}$$

(2) Translational motion estimation from a motion blurred image and an unblurred image by using central geometric moments

From Eqs. (22) and (23), we see that the displacement $\mathbf{d}_0$ of the object in the motion blurred image at time $t_s^{(i)}$ with respect its position in the unblurred reference image must be calculated when estimating the parameter $\mathbf{Para}$ matrix of the motion. But in many practical applications, we need only estimate $\mathbf{Para}$ and don't care $\mathbf{d}_0$. In this case, central moments are used to realize motion estimation.

Obviously, the central moments of the motion are irrelative to the displacement $\mathbf{d}_0$, as is shown in Eq. (24):

$$\mu_{pq}^{\left(d^{(i)}\right)} = \frac{1}{t_e^{\left(g^{(i)}\right)}} \int_{t_s^{(i)}}^{t_s^{(i)}+t_e^{\left(g^{(i)}\right)}} \left(d_x(t)-\overline{d_x}\right)^p \left(d_y(t)-\overline{d_y}\right)^q dt$$
$$= \frac{1}{t_e^{\left(g^{(i)}\right)}} \int_{t_s^{(i)}}^{t_s^{(i)}+t_e^{\left(g^{(i)}\right)}} \left(d_x(t)-d_{x0}-\overline{d_x^{/}}\right)^p \left(d_y(t)-d_{y0}-\overline{d_y^{/}}\right)^q dt \tag{24}$$

where $\overline{d_x^{/}} = \frac{1}{t_e^{\left(g^{(i)}\right)}} \int_{t_s^{(i)}}^{t_s^{(i)}+t_e^{\left(g^{(i)}\right)}} \left(d_x(t)-d_{x0}\right)dt, \overline{d_y^{/}} = \frac{1}{t_e^{\left(g^{(i)}\right)}} \int_{t_s^{(i)}}^{t_s^{(i)}+t_e^{\left(g^{(i)}\right)}} \left(d_y(t)-d_{y0}\right)dt$.

Therefore, according to Theorem 3, we can employ central moments to calculate the motion parameter matrix **Para** by solving the following nonlinear equations:

$$\Phi_{pq}^{*}\left(\mathbf{Para}\right) - \mu_{pq}^{\left(d^{(i)}\right)} = 0, \left(0 \le p \le m, 0 \le q \le n\right) \tag{25}$$

where Φ_{pq}^{*} is a nonlinear function of **Para**; $m \ge N_x$ and $N \ge N_y$; $\mu_{pq}^{\left(d^{(i)}\right)}$ is given by:

$$\mu_{pq}^{\left(d^{(i)}\right)} = \left[\frac{\rho^{(f)} t_e^{(f)}}{\rho^{\left(g^{(i)}\right)} t_e^{\left(g^{(i)}\right)}} \mu_{pq}^{\left(g^{(i)}\right)} - \sum_{k=0}^{p-1}\sum_{l=0}^{q-1} \binom{p}{k}\binom{q}{l} \mu_{kl}^{\left(d^{(i)}\right)} \mu_{p-k,q-l}^{(f)} - \sum_{k=0}^{p-1} \binom{p}{k} \mu_{kq}^{\left(d^{(i)}\right)} \mu_{p-k,0}^{(f)} \right.$$
$$\left. - \sum_{l=0}^{q-1} \binom{q}{l} \mu_{pl,}^{\left(d^{(i)}\right)} \mu_{0,q-l}^{(f)} \right] / \mu_{00}^{(f)} \tag{26}$$

Similarly to Eq. (23), Eqs.(25) can be solved by:

$$\left\{\hat{\mathbf{Para}}\right\} = \arg\min_{\mathbf{Para}} \varepsilon\left(\mathbf{Para}\right) = \arg\min_{\mathbf{Para}} \left(\sum_{p=0}^{m}\sum_{q=0}^{n} \left\| \Phi_{pq}^{*}\left(\mathbf{Para}\right) - \mu_{pq}^{\left(d^{(i)}\right)} \right\|^2 \right) \tag{27}$$

(3) Translational motion estimation from two successive frames of motion blurred images by using geometric moments

We can also use two successive blurred frames $g^{(i)}$ and $g^{(i+1)}$ to calculate the motion parameters. In this case, we hypothesize the elements of **Para** keep constant in the time interval $\left[t_s^{(i)}, t_s^{(i+1)} + t_e^{\left(g^{(i+1)}\right)}\right]$, $\mathbf{Para} = \hat{\mathbf{Para}}$. The motion moments

$m_{pq}^{\left(d^{(i)}\right)}(\mathbf{d}_0)$ and $m_{pq}^{\left(d^{(i+1)}\right)}(\mathbf{d}_0)$ with respect to the $\mathbf{d}_0$ within $\left[t_s^{(i)}, t_s^{(i)}+t_e^{\left(g^{(i)}\right)}\right]$ and $\left[t_s^{(i+1)}, t_s^{(i+1)}+t_e^{\left(g^{(i+1)}\right)}\right]$ are determined by $\mathbf{d}(t)$ or the assumed parameter matrix $\hat{\mathbf{Para}}$, as is shown in Eq. (27):

$$\begin{aligned} m_{pq}^{\left(d^{(i+j)}\right)}(\mathbf{d}_0) &= \frac{1}{t_e^{\left(g^{(i+j)}\right)}} \int_{t_s^{(i+j)}}^{t_s^{(i+j)}+t_e^{\left(g^{(i+j)}\right)}} \left(d_x(t)-d_{x0}\right)^p \left(d_x(t)-d_{y0}\right)^q dt \\ &= \frac{1}{t_e^{\left(g^{(i+j)}\right)}} \int_{t_s^{(i+j)}}^{t_s^{(i+j)}+t_e^{\left(g^{(i+j)}\right)}} \left(\Gamma_x\left(\hat{\mathbf{Para}}_x, \left[(t-t_i)^0, \cdots, (t-t_i)^{N-1}\right]^T\right)\right)^p \\ &\times \left(\Gamma_y\left(\hat{\mathbf{Para}}_y, \left[(t-t_i)^0, \cdots, (t-t_i)^{N-1}\right]^T\right)\right)^q dt \end{aligned} \tag{28}$$

Thus, we can calculate two values $M_{pq}^{(f_0^*)}$ and $M_{pq}^{(f_1^*)}$ for the $(p+q)$-th order moment of the unblurred image $f^*(x,y)=f(x-d_{x0}, y-d_{y0})$ from $g^{(i)}$ and $g^{(i+1)}$ using Theorem 2, respectively. The iterative formula for computing $M_{pq}^{(f_0^*)}$ and $M_{pq}^{(f_1^*)}$ from the moments of the assumed motion and the moments of the two motion blurred images is given by:

$$\begin{aligned} M_{pq}^{\left(f_j^*\right)} = \Bigg[& \frac{t_e^{\left(f^*\right)}}{t_e^{\left(g^{(i+j)}\right)}} M_{pq}^{\left(g^{(i+j)}\right)} - \sum_{k=0}^{p-1}\sum_{l=0}^{q-1} \binom{p}{k}\binom{q}{l} M_{kl}^{\left(f_j^*\right)} m_{p-k,q-l}^{\left(d^{(i+j)}\right)}(\mathbf{d}_0) \\ & - \sum_{k=0}^{p-1} \binom{p}{k} M_{kq}^{\left(f_j^*\right)} m_{p-k,0}^{\left(d^{(i+j)}\right)}(\mathbf{d}_0) - \sum_{l=0}^{q-1} \binom{q}{l} M_{pl,}^{\left(f_j^*\right)} m_{0,q-l}^{\left(d^{(i+j)}\right)}(\mathbf{d}_0) \Bigg] / m_{00}^{\left(d^{(i+j)}\right)}(\mathbf{d}_0) \end{aligned} \tag{29}$$

where $j=0,1$.

When the hypothetical motion parameter matrix is correct, the two values $M_{pq}^{(f_0^*)}$ and $M_{pq}^{(f_1^*)}$ will be identical for any p and q, otherwise the error between $M_{pq}^{(f_0^*)}$ and $M_{pq}^{(f_1^*)}$ will be large. Thus, the true motion parameter matrix $\mathbf{Para}$ can be found by searching for the minimum of the error:

$$\hat{\mathbf{Para}} = \arg\min_{\mathbf{Para}} \varepsilon(\mathbf{Para}) = \arg\min_{\mathbf{Para}} \left(\sum_{p=0}^{m}\sum_{q=0}^{n} \left\| M_{pq}^{(f_0^*)} - M_{pq}^{(f_1^*)} \right\|^2 \right). \tag{30}$$

3.1.1.3. Applications of the Algorithms for Translational Motion Estimation Using Geometric Moments

The algorithms for estimating translational motion based on the spatial model of motion blurred images and geometric moments can be applied to motions of various forms, such as uniform motion, accelerated motion, vibrations, etc.

3.1.1.3.1. Uniform Motion

For 2D in-plane uniform motion, the velocity $\mathbf{v}$ of the object (or image motion) keeps uniform within the time interval $\left[t_s^{(i)}, t_s^{(i+1)} + t_e^{\left(g^{(i+1)}\right)}\right]$, the motion can be expressed by:

$$\begin{bmatrix} d_x(t) \\ d_y(t) \end{bmatrix} = \begin{bmatrix} d_x\left(t_s^{(i)}\right) & v_x \\ d_y\left(t_s^{(i)}\right) & v_y \end{bmatrix} \begin{bmatrix} 1 \\ t - t_i \end{bmatrix}, t \in \left[t_s^{(i)}, t_s^{(i+1)} + t_e^{\left(g^{(i+1)}\right)}\right] \tag{31}$$

where $(d_x(t_s^{(i)}), d_y(t_s^{(i)}))^T$ is the displacement vector of the object in the motion blurred image at time $t_s^{(i)}$ with respect its position in the unblurred reference image, v_x and v_y denote the image velocity components of $\mathbf{v}$ in x and y directions respectively.

From Eq. (14), (15) and Eq. (30), we get the $(p+0)$-th and $(0+q)$-th order geometric moments of the motion within the time interval $\left[t_s^{(i)}, t_s^{(i)} + t_e^{\left(g^{(i)}\right)}\right]$ which are given by:

$$m_{p0}^{\left(d^{(i)}\right)}(\mathbf{0}) = \frac{1}{t_e^{\left(g^{(i)}\right)}} \int_{t_s^{(i)}}^{t_s^{(i)} + t_e^{\left(g^{(i)}\right)}} \left(d_x\left(t_s^{(i)}\right) + v_x\left(t - t_s^{(i)}\right)\right)^p dt = \sum_{k=0}^{p} \binom{p}{k} \lambda_k^{(i)} v_x^k d_x\left(t_s^{(i)}\right)^{p-k} \tag{32}$$

$$m_{0q}^{\left(d^{(i)}\right)}(\mathbf{0}) = \sum_{k=0}^{q} \binom{q}{k} \lambda_k^{(i)} v_y^k d_y\left(t_s^{(i)}\right)^{q-k} \tag{33}$$

The $(p+0)$-th and $(0+q)$-th order central moments of the motion within $\left[t_s^{(i)}, t_s^{(i)} + t_e^{\left(g^{(i)}\right)}\right]$ are written as:

$$\mu_{p0}^{\left(d^{(i)}\right)} = \frac{1}{t_e^{\left(g^{(i)}\right)}} \int_{t_s^{(i)}}^{t_s^{(i)}+t_e^{\left(g^{(i)}\right)}} \left(d_x\left(t_s^{(i)}\right) + v_x\left(t - t_s^{(i)}\right) - \overline{d_x^{(i)}}\right)^p dt$$
$$= \frac{v_x^p}{(p+1)t_e^{\left(g^{(i)}\right)}} \left(\left(t_e^{\left(g^{(i)}\right)}/2\right)^{p+1} - \left(-t_e^{\left(g^{(i)}\right)}/2\right)^{p+1}\right) \quad (34)$$

$$\mu_{0q}^{\left(d^{(i)}\right)} = \frac{v_y^q}{(q+1)t_e^{\left(g^{(i)}\right)}} \left(\left(t_e^{\left(g^{(i)}\right)}/2\right)^{q+1} - \left(-t_e^{\left(g^{(i)}\right)}/2\right)^{q+1}\right) \quad (35)$$

The $(p+0)$-th and $(0+q)$-th order moments of the motion within $\left[t_s^{(i)}, t_s^{(i)} + t_e^{\left(g^{(i)}\right)}\right]$ and $\left[t_s^{(i+1)}, t_s^{(i+1)} + t_e^{\left(g^{(i+1)}\right)}\right]$ with respect to the displacement $\mathbf{d}\left(t_s^{(i)}\right)$ are given by:

$$m_{p0}^{\left(d^{(i+j)}\right)}\left(\mathbf{d}\left(t_s^{(i)}\right)\right) = \frac{1}{t_e^{\left(g^{(i+j)}\right)}} \int_{t_s^{(i+j)}}^{t_s^{(i+j)}+t_e^{\left(g^{(i+j)}\right)}} \left(d_x\left(t_s^{(i)}\right) + v_x\left(t - t_s^{(i)}\right) - d_x\left(t_s^{(i)}\right)\right)^p dt = \lambda_p^{(i+j)} v_x^p \quad (36)$$

.

$$m_{0q}^{\left(d^{(i+j)}\right)}\left(\mathbf{d}\left(t_s^{(i)}\right)\right) = \lambda_q^{(i+j)} v_y^q \quad (37)$$

where $j = 0,1$; $\lambda_k^{(i+j)} = \dfrac{\left(t_s^{(i+j)} + t_e^{\left(g^{(i+j)}\right)} - t_s^{(i)}\right)^{k+1} - \left(t_s^{(i+j)} - t_s^{(i)}\right)^{k+1}}{t_e^{\left(g^{(i+j)}\right)}(k+1)}$, k is any positive integer.

Eqs. (29-34) show that $m_{p0}^{\left(d^{(i)}\right)}(\mathbf{0})$ is a function of $\left(d_x\left(t_s^{(i)}\right), v_x\right)^T$, $m_{0q}^{\left(d^{(i)}\right)}(\mathbf{0})$ is a function of $\left(d_y\left(t_s^{(i)}\right), v_y\right)^T$, $\mu_{p0}^{\left(d^{(i)}\right)}$ and $m_{p0}^{\left(d^{(i+j)}\right)}\left(\mathbf{d}\left(t_s^{(i)}\right)\right)$ are functions of v_x, $\mu_{0q}^{\left(d^{(i)}\right)}$ and $m_{0q}^{\left(d^{(i+j)}\right)}\left(\mathbf{d}\left(t_s^{(i)}\right)\right)$ are functions of v_y. Thus, the two velocity component v_x and v_y can be calculated from an unblurred image and a motion blurred image by using Eqs. (29-30), Eq. (19) and Eq. (23) or using Eqs. (31-32), Eq.(26) and Eq. (27). We can also estimate v_x and v_y from two successive motion blurred images by employing Eqs. (33-34), Eq. (29) and Eq. (30).

3.1.1.3.2. Accelerated Motion

Accelerated motion can also be estimated using our algorithm. For example, if an image motion is assumed to have an initial velocity $\mathbf{v}_0$ at time $t_s^{(i)}$ and a uniform acceleration $\mathbf{a}$, then the motion can be expressed by:

$$\begin{bmatrix} d_x(t) \\ d_y(t) \end{bmatrix} = \begin{bmatrix} d_x\left(t_s^{(i)}\right) & v_{x0} & a_x \\ d_y(t_i) & v_{y0} & a_y \end{bmatrix} \begin{bmatrix} 1 \\ t-t_s^{(i)} \\ \left(t-t_s^{(i)}\right)^2/2 \end{bmatrix}, t \in \left[t_s^{(i)}, t_s^{(i+1)} + t_e^{\left(g^{(i+1)}\right)}\right] \quad (38)$$

where $(d_x(t_s^{(i)}), d_y(t_s^{(i)}))^T$ is the displacement vector of the object in the motion blurred image with respect to its position in the unblurred image $t_s^{(i)}$, $\mathbf{v} = (v_{x0}, v_{y0})^T$ is the initial image velocity at time $t_s^{(i)}$, $(a_x, a_y)^T$ is the acceleration vector.

According to Eq. (14), Eq. (15) and Eq. (38), the moments with respect to $\mathbf{0}$, the central moments, the moments with respect to $\mathbf{d}\left(t_s^{(i)}\right)$ of the accelerated motion can be given by:

$$\begin{aligned} m_{p0}^{\left(d^{(i)}\right)}(\mathbf{0}) &= \frac{1}{t_e^{\left(g^{(i)}\right)}} \int_{t_s^{(i)}}^{t_s^{(i)}+t_e^{\left(g^{(i)}\right)}} \left(d_x\left(t_s^{(i)}\right) + v_{x0}\left(t-t_s^{(i)}\right) + a_x\left(t-t_s^{(i)}\right)^2/2\right)^p dt \\ &= \sum_{k=0}^{p} \binom{p}{k} d_x\left(t_s^{(i)}\right)^{p-k} \left(\sum_{l=0}^{k} \binom{k}{l} \lambda_{2k-l}^{(i)} v_{x0}^l \left(\frac{a_x}{2}\right)^{k-l}\right) \end{aligned} \quad (39)$$

$$m_{0q}^{\left(d^{(i)}\right)}(\mathbf{0}) = \sum_{k=0}^{q} \binom{q}{k} d_y\left(t_s^{(i)}\right)^{q-k} \left(\sum_{l=0}^{k} \binom{k}{l} \lambda_{2k-l}^{(i)} v_{y0}^l \left(\frac{a_y}{2}\right)^{k-l}\right) \quad (40)$$

$$\begin{aligned} \mu_{p0}^{\left(d^{(i)}\right)} &= \frac{1}{t_e^{\left(g^{(i)}\right)}} \int_{t_s^{(i)}}^{t_s^{(i)}+t_e^{\left(g^{(i)}\right)}} \left(d_x\left(t_s^{(i)}\right) + v_{x0}\left(t-t_s^{(i)}\right) + a_x\left(t-t_s^{(i)}\right)^2/2 - \overline{d_x^{(i)}}\right)^p dt \\ &= \sum_{k=0}^{p} \binom{p}{k} v_{x0}^k a_{x0}^{p-k} \xi_k^{(i)} \end{aligned} \quad (41)$$

$$\mu_{0q}^{\left(d^{(i)}\right)} = \sum_{k=0}^{q} \binom{q}{k} v_{y0}^k a_{y0}^{q-k} \xi_k^{(i)} \quad (42)$$

where $\xi_k^{(i)} = \left(\left(\frac{t_e^{(g^{(i)})}}{2} \right)^k \left(t_e^{(g^{(i)})} - \frac{\left(t_e^{(g^{(i)})}\right)^2}{6} \right)^{p-k} - \left(-\frac{t_e^{(g^{(i)})}}{2} \right)^k \left(-\frac{\left(t_e^{(g^{(i)})}\right)^2}{6} \right)^{p-k} \right)$.

$$m_{p0}^{\left(d^{(i+j)}\right)}\left(\mathbf{d}\left(t_s^{(i)}\right)\right) = \frac{1}{t_e^{\left(g^{(i+j)}\right)}} \int_{t_s^{(i)}}^{t_s^{(i+j)}+t_e^{\left(g^{(i+j)}\right)}} \left(v_{x0}\left(t-t_s^{(i)}\right) + a_x\left(t-t_s^{(i)}\right)^2/2 \right)^p dt$$
$$= \sum_{k=0}^{p} \binom{p}{k} \lambda_{2p-k}^{(i+j)} v_{x0}^k \left(\frac{a_x}{2}\right)^{p-k} \tag{43}$$

$$m_{0q}^{\left(d^{(i+j)}\right)}\left(\mathbf{d}\left(t_s^{(i)}\right)\right) = \sum_{k=0}^{q} \binom{q}{k} \lambda_{2q-k}^{(i+j)} v_{y0}^k \left(\frac{a_y}{2}\right)^{q-k} \tag{44}$$

Theoretically, the initial velocity $\mathbf{v}_0$ and the acceleration $\mathbf{a}$ can be computed using Eqs. (36-37), Eq. (19) and Eq. (23) or using Eqs. (38-39), Eq.(26) and Eq. (27) from a motion blurred image and an unblurred image or using Eqs. (40-41), Eq. (29) and Eq. (30) from two successive motion blurred images. But in practical situations, the "motion blur" information of one blurred image is not sufficient to produce credible results. Therefore, we use two successive blurred images to estimate an accelerated motion by Eqs. (40-41), Eq. (29) and Eq. (30).

3.1.1.3.3. Sinusoidal Vibration

The proposed algorithm can also be applied to vibrations. In this chapter, we introduce a simple sample, i.e. sinusoidal vibration, to describe its implementation in measuring vibrations.

For sinusoidal translational vibrations, the displacement function of the image is given by:

$$\mathbf{d}(t) = \begin{bmatrix} d_x(t) \\ d_y(t) \end{bmatrix} = \begin{bmatrix} \overline{d_x} \\ \overline{d_y} \end{bmatrix} + \begin{bmatrix} A\cos\left(2\pi f_V\left(t-t_s^{(i)}\right)+\phi_s^{(i)}\right)\cos\theta \\ A\cos\left(2\pi f_V\left(t-t_s^{(i)}\right)+\phi_s^{(i)}\right)\sin\theta \end{bmatrix}, \tag{45}$$

where A and f_V are the amplitude and frequency of the vibration, respectively; $\begin{bmatrix}\overline{d_x} & \overline{d_y}\end{bmatrix}^T$ is the average displacement of the object in the motion blurred image with respect to its initial position in the unblurred image; $\phi_s^{(i)}$ is the phase of the vibration at time $t_s^{(i)}$; θ is the angle between the direction of the vibration and the x-axis of the image plane.

From Eq. (14) and Eq. (45), we get the (*p* + *q*)-th order moment $m_{pq}^{(Vib)}$ of the vibration:

$$m_{pq}^{\left(V^{(i+j)}\right)} = \frac{1}{t_e^{\left(g^{(i+j)}\right)}} \int_{t_s^{(i+j)}}^{t_s^{(i+j)}+t_e^{\left(g^{(i+j)}\right)}} \left(\overline{d_x} + A\cos\left(2\pi f_V\left(t-t_s^{(i)}\right)+\phi_s^{(i)}\right)\cos\theta\right)^p \times\left(\overline{d_y} + A\cos\left(2\pi f_V\left(t-t_s^{(i)}\right)+\phi_s^{(i)}\right)\sin\theta\right)^q dt \tag{46}$$

According to *binomial theorem*, Eq. (46) can be written by:

$$m_{pq}^{\left(V^{(i+j)}\right)} = \frac{1}{t_e^{\left(g^{(i+j)}\right)}} \sum_{k=0}^{p}\sum_{l=0}^{q}\binom{p}{k}\binom{q}{l}\overline{d_x}^k\,\overline{d_y}^l A^{(p-k)+(q-l)} \times\cos^{p-k}\theta\sin^{q-l}\theta F_{(p-k)+(q-l)}\left(t_s^{(i+j)}, t_s^{(i)}, t_e^{\left(g^{(i+j)}\right)}, f_V, \phi_s^{(i)}\right) \tag{47}$$

where

$$F_k\left(t_s^{(i+j)}, t_s^{(i)}, t_e^{\left(g^{(i+j)}\right)}, f_V, \phi_s^{(i)}\right) = \int_{t_s^{(i+j)}}^{t_s^{(i+j)}+t_e^{\left(g^{(i+j)}\right)}} \cos\left(2\pi f_V\left(t-t_s^{(i)}\right)+\phi_s^{(i)}\right)^k dt,$$

F_k can be calculated iteratively by:

$$F_0 = t_e^{\left(g^{(i+j)}\right)}$$

$$F_1 = \frac{\sin\left(2\pi f_V\left(t_s^{(i+j)}+t_e^{\left(g^{(i+j)}\right)}-t_s^{(i)}\right)+\phi_s^{(i)}\right)-\sin\left(2\pi f_V\left(t_s^{(i+j)}-t_s^{(i)}\right)+\phi_s^{(i)}\right)}{2\pi f_V}$$

$$F_k = \frac{\cos^{k-1}\left(2\pi f_V\left(t_s^{(i+j)}+t_e^{\left(g^{(i+j)}\right)}-t_s^{(i)}\right)+\phi_s^{(i)}\right)\sin\left(2\pi f_V\left(t_s^{(i+j)}+t_e^{\left(g^{(i+j)}\right)}-t_s^{(i)}\right)+\phi_s^{(i)}\right)}{2k\pi f_V} - \frac{\cos^{k-1}\left(2\pi f_V\left(t_s^{(i+j)}-t_s^{(i)}\right)+\phi_s^{(i)}\right)\sin\left(2\pi f_V\left(t_s^{(i+j)}-t_s^{(i)}\right)+\phi_s^{(i)}\right)}{2k\pi f_V}+\frac{k-1}{k}F_{k-2}$$

$$for \quad k\geq 2, and \quad j=\cdots,-2,-1,0,1,2,\cdots \tag{48}$$

From Eq. (47), we see that $m_{pq}^{(Vib)}$ is a function of the amplitude A, the frequency f_V, the phase $\phi_s^{(i)}$ at time $t_s^{(i)}$, the direction ψ of the vibration and the average displacement $\bar{\mathbf{d}}$. Therefore, we can estimate A, f_V, $\phi_s^{(i)}$ ψ by using the proposed algorithm if an unblurred image and a motion blurred image are acquired or if we capture two successive motion blurred images.

3.1.1.4. Experimental Results of the Geometric-Moment-Based Algorithms for Translational Motion Estimation

In this section, we designed corresponding experiments to validate the performance of the spatial-model-based algorithms for various applications.

3.1.1.4.1. Uniform Motion

Experiment 1: The principle for estimating 2D translational motions is quite similar to that for 1D ones, therefore, we synthesize two pairs of successive motion blurred images with uniform motion to measure the performance of the proposed algorithm in uniform motion estimation. One pair is with a velocity of 10 pixel/$t_e^{(g)}$ in x direction, where $t_e^{(g)}$ is the exposure time of the motion blurred images, and is equal to the exposure time $t_e^{(f)}$ of the unblurred reference image. The other is with a velocity of 20 pixel/$t_e^{(g)}$ in y direction. The y - velocity components of both groups are set to zero. The simulated temporal sampling period for synthesize both groups is $2t_e^{(g)}$. Fig.2a gives the unblurred reference image we

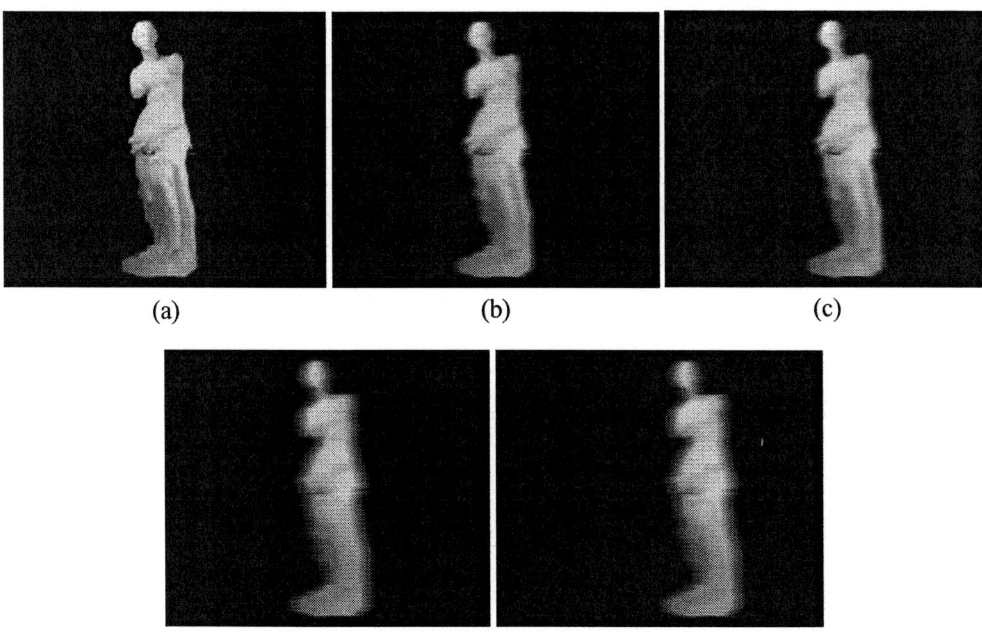

(a) (b) (c)

(d) (e)

Figure 2. (a) The unblurred reference image; (b) and (c) The successive motion blurred images with the velocity of 10 pixel/$t_e^{(g)}$; (d) and (e) The successive motion blurred images with the velocity of 20 pixel/$t_e^{(g)}$;

used to generate the simulated motion blurred images. Fig.2b and Fig.2c show the motion blurred image pair with the velocity of 10 pixel/$t_e^{(g)}$. Fig.2d and Fig.2e are motion blurred images with the velocity of 20 pixel/$t_e^{(g)}$.

The experimental results of Experiment 1 given in Tab.1a-c proves that our algorithm has good accuracy for estimating uniform motions.

Table 1a. The experimental results for estimating uniform motion from a motion blurred image and an unblurred reference image by using geometric moments*

The images used in this algorithm	v^c (in pixel/$t_e^{(g)}$)	d^c (in pixel)	v^e (in pixel/$t_e^{(g)}$)	d^e (in pixel)	ε^v (%)
Fig.2a and Fig.2b	10	0	10.2225	-0.1104	2.225
Fig.2a and Fig.2d	20	0	20.2466	-0.1234	1.233

Table 1b. The experimental results for estimating uniform motion from a motion blurred image and an unblurred reference image by using central moments*

The images used in this algorithm	v^c (in pixel/$t_e^{(g)}$)	v^e (in pixel/$t_e^{(g)}$)	ε^v (%)
Fig.2a and Fig.2c	10	10.1922	1.922
Fig.2a and Fig.2e	20	20.2476	1.238

Table 1c. The experimental results for estimating uniform motion from two successive frames of motion blurred images by using geometric moments*

The images used in this algorithm	v^c (in pixel/$t_e^{(g)}$)	v^e (in pixel/$t_e^{(g)}$)	ε^v (%)
Fig.2b and Fig.2c	10	10.0001	0.001
Fig.2d and Fig.2e	20	20.0000	0.000

* v^c and d^c denote the correct velocities of the motion blurred images and the displacements of the motion blurred images relative to the unblurred reference image, respectively. v^e and d^e are the corresponding estimated values of the velocities and the displacements, respectively, ε^v is the relative estimated error of velocities.

3.1.1.4.2. Accelerated Motion

Experiment 2: In this experiment, we verify the accuracy of the presented algorithm for measuring accelerated motion. Two pairs of synthetic motion blurred images with different initial velocities and accelerations are provided in Fig.3a-b and Fig.3c-d, respectively. The first pair (shown in Fig.3a-b) is with the initial velocity of 10pixel/$t_e^{(g)}$ and the acceleration of

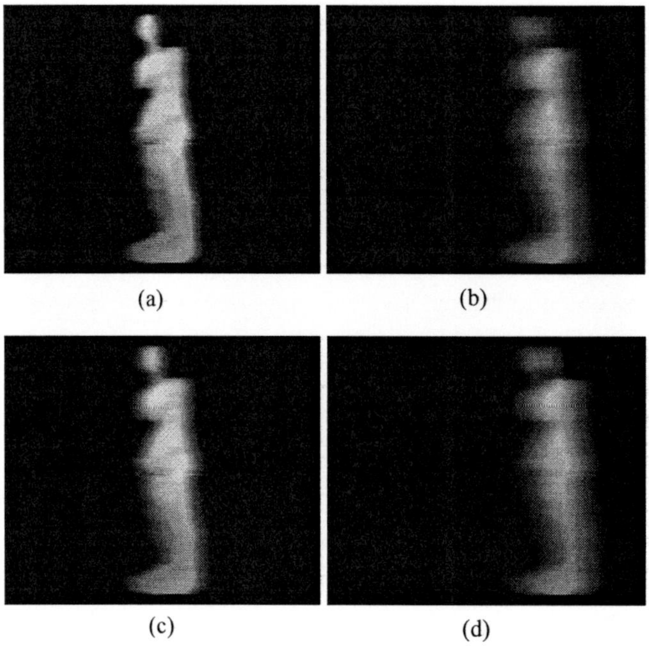

Figure 3. (a) and (b) The successive motion blurred images with the initial velocity of 10 pixel/$t_e^{(g)}$ and the acceleration of 15 pixel/$t_e^{(g)2}$; (c) and (d) The successive motion blurred images with the velocity of 20 pixel/$t_e^{(g)}$ and the acceleration of 20 pixel/$t_e^{(g)2}$.

15 pixel/$t_e^{(g)2}$. The second one is with the initial velocity of 20pixel/$t_e^{(g)}$ and the acceleration of 10 pixel/$t_e^{(g)2}$, as is shown in Fig.3c-d. The simulated exposure time and the temporal sampling period are also supposed to be $t_e^{(g)}$ and $2t_e^{(g)}$, where $t_e^{(g)} = t_e^{(f)}$.

The experimental results of Experiment 2 are provided in Tab.2. From Tab.2, we see that our algorithm is quite effective when it is applied to accelerated motions.

Table 2. The experimental results for estimating accelerated motion from two successive frames of motion blurred images by using geometric moments*

The images used in this algorithm	v_0^c **(in pixel/**$t_e^{(g)}$**)**	a^c **(in pixel/**$t_e^{(g)2}$**)**	v_0^e **(in pixel/**$t_e^{(g)}$**)**	a^e **(in pixel/**$t_e^{(g)2}$**)**	ε^{v_0} **(%)**	ε^a **(%)**
Fig.3a and Fig.3b	10	15	9.5442	15.3044	4.5580	2.029
Fig.3a and Fig.3d	20	10	19.7023	10.1984	1.4885	1.984

* v_0^c and a^c denote the correct initial velocities and the acceleration of the motion blurred images, respectively. v_0^e and a^e are the corresponding estimated values of the initial velocities and the accelerations, respectively, ε^v and ε^a is the relative estimated error of velocities and accelerations, respectively.

3.1.1.4.3. Sinusoidal Vibration

Experiment 3: In this section, we validate the performance of the proposed algorithms for sinusoidal vibration measurement. We designed an experimental platform to generate real sinusoidal translational vibrations and capture motion blurred images of the vibrating object for testing the performance of the presented algorithm in vibration measurement. This platform uses a vibration exciter to generate vibrations with various amplitudes and frequencies. The directions of the vibrations are approximately orthogonal to the optical axis of the camera and parallel to the x-axis of the image plane of the camera. An object with a surface parallel to the image plane is jointed with the output side of the vibration exciter. We acquires sequences of motion blurred images by A BASLER camera, which exposure time and sampling period are set to be 36ms and 72ms, respectively. The sizes of the acquired images are 656×491. The proportion coefficient between the length unit (in pixel) of the images captured by this camera and the physical length unit (in mm) is 0.027848mm/pixel. The schematics of the experimental platform and part of the real experiment equipment are shown in Fig.4a and Fig.4b, respectively.

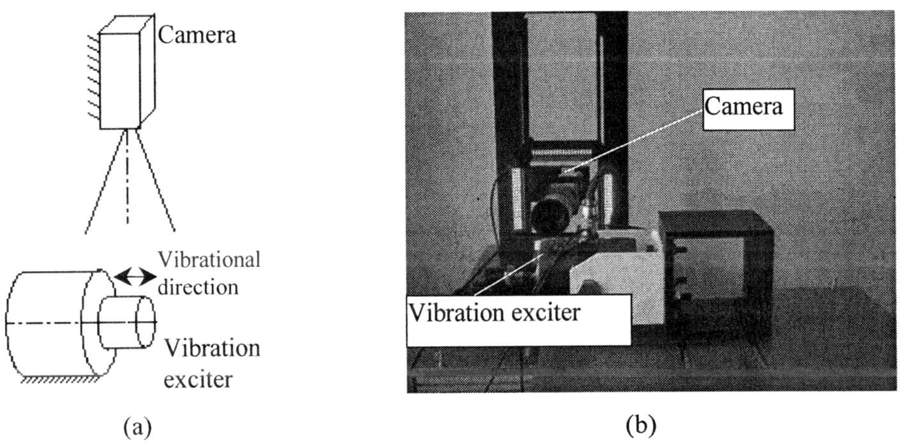

Figure 4. (a) schematics of the experimental platform in Experiment 3; (b) part of the real experiment equipment.

This experiment measures the vibrational parameters by using two frames of successive motion blurred images. Two pairs of successive motion blurred image frames with different amplitudes and frequencies are employed to verify our algorithm. The amplitude and frequency of the object keep constant when catching each image. The vibrational amplitude and frequency of the object are approximately 0.39mm and 8.3Hz when acquiring image pair 1 shown in Fig.5a and Fig.5b. And when capturing image pair 2 shown in Fig.5c-d, the vibrational amplitude and frequency are approximately 0.75mm and 6Hz respectively.

The experimental results of Experiment 3 are given in Tab.3. The measured results from an accelerometer are also provided in Tab.3 for comparison. According to Tab.3, we know that our algorithm has nice performance in translational vibration measurement.

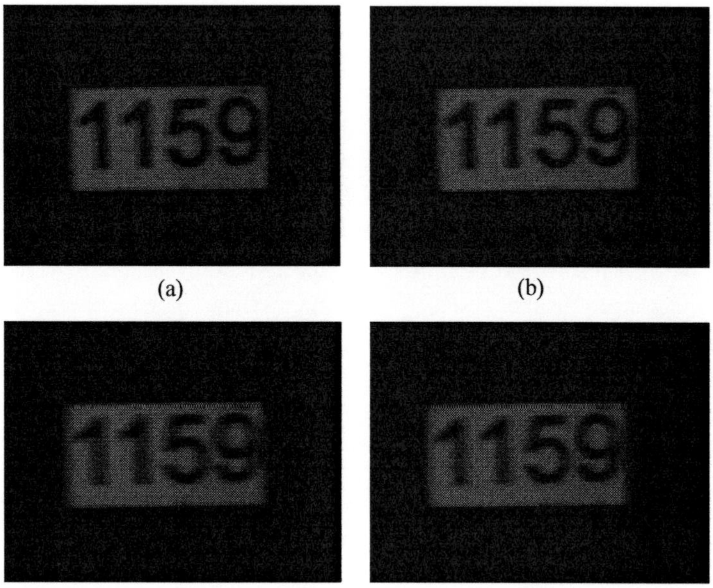

(a) (b) (c) (d)

Figure 5. (a) and (b) Successive motion blurred image pair 1: when acquiring it, the vibrational amplitude and frequency of the object are 0.39mm and 8.3Hz; (c) and (d) Successive motion blurred image pair 2: when acquiring it, the vibrational amplitude and frequency of the object are 0.75mm and 6Hz;

Table 3. The experimental results for measuring sinusoidal vibrations by using geometric moments*

The images used in this algorithm	A^a (in mm)	f_V^a (in Hz)	A^e (in mm)	f_V^e (in Hz)
Fig.5a and Fig.5b	0.39mm	8.3	0.3816	15.3044
Fig.5c and Fig.5d	0.75mm	6.1	0.7673	6.317

* A^a and f_V^a are the values measured by the accelerometer; A^e and f_V^e are the corresponding values measured by the proposed algorithm.

3.1.2. Estimating Monaxial Rotations by Using Geometric Moments

3.1.2.1. Theory of Monaxial Rotation Estimation by Using Geometric Moments

In industrial applications, monaxial rotation is another widespread form of motion. According to the spatial model of the motion blurred images for monaxial rotations described in section 2.2.3, we know that estimation of a monaxial rotation is equivalent to the measurement of a 1D translation if we substitute a polar coordinate system for the Cartesian coordinate system which is used in section 3.1.1 for estimating translational motions.

However, for polar images (shown in section 2.1.2.3), the polar angle is infinite and the positive infinity $+\infty$ of the polar angle is jointed to the negative infinity $-\infty$, which causes difficulties in applying the approach proposed in section 3.1.1 directly to the estimation of monaxial rotations. To avoid this, we at first set a "white against black" mark on the object surface. The white region of the mark has no intersection with the rotational axis and is surrounded by the black region, which occupies the rest view of the camera. For the black regions in the images, the illuminant is much less than the white regions. Thus, we can suppose the illuminant of the black region is zero. Further, we suppose the loci of the white region of the mark don't form a whole circle (as is shown in fig.6), i.e., the angular displacement of the object within the time interval $\left[t_s^{(i)}, t_s^{(i+1)} + t_e^{g^{(i+1)}}\right]$ satisfies $\Delta\phi_{\max} - \Delta\phi_{\min} < 2\pi - (\phi_{\max} - \phi_{\min})$, where $\phi_{\max}$ and $\phi_{\min}$ are the maximum and the minimum polar angles for the white region of the mark in the unblurred reference image, respectively; $\Delta\phi_{\max}$ and $\Delta\phi_{\min}$ ($-2\pi < \Delta\phi_{\min} < \phi_{\max} - \phi_{\min}$) denote the maximum and minimum angular displacements of the object, respectively. Then we can restrict the integral interval of the polar angle to a certain finite integral $[\phi_L, \phi_L + 2\pi]$ when calculating the geometric moment of the polar images, as is shown in Eq. (49) and Eq. (50):

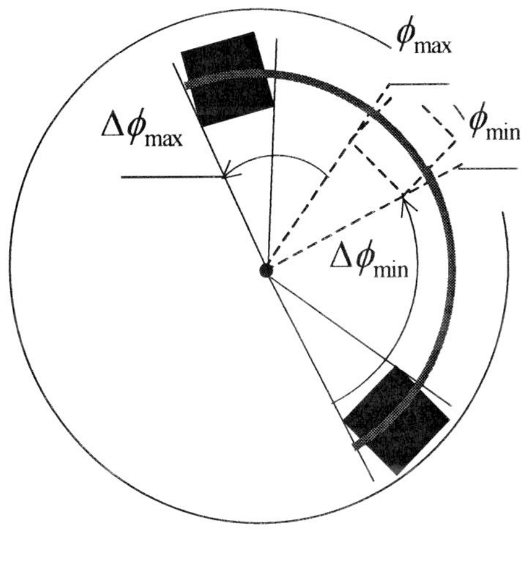

Figure 6. The angular displacement of the bright region.

$$M_{pq}^{(f)} = \int_{r_{\min}}^{r_{\max}} \int_{\phi_L}^{\phi_L + 2\pi} r^p \phi^q f(r,\phi) d\phi dr \tag{49}$$

$$M_{pq}^{\left(g^{(i)}\right)} = \int_{r_{\min}}^{r_{\max}} \int_{\phi_L}^{\phi_L} r^p \phi^q g^{(i)}(r,\phi) d\phi dr \tag{50}$$

where $\phi_L = \min(\phi_{\min} + \Delta\phi_{\min}, \phi_{\min})$. ϕ_L can be determined easily by image projection, as is described as follows:

For motion blurred images $g^{(i)}$, when the image coordinates (r,ϕ) satisfy $r \in \left[R_{\min}^w, R_{\max}^w\right]$ and $\phi \in [\phi_{\min} + \Delta\phi_{\min}, \phi_{\max} + \Delta\phi_{\max}]$, $g^{(i)} \geq T_b$; otherwise, $g^{(i+j)} = T_b \approx 0$, where $R_{\min}^w$ and $R_{\max}^w$ are the minimum and maximum polar radius of the periphery of the white region of the mark, respectively; T_b is the mean illuminant of the black region in the images. For the unblurred reference image, when (r,ϕ) satisfies $r \in \left[R^w_{\min}, R^w_{\max}\right]$ and $\phi \in \left[\phi_{\min}, \phi_{\max}\right]$, $f \geq T_b$; otherwise, $f = T_b$. Therefore, we let the "combined image" $G = g^{(i)} + f$ or $G = \sum_j g^{(i+j)}$, and project G onto the θ-axis to determine ϕ_L. The corner whose θ-coordinate is the minimum among all corners in the projection is $\phi_L = \min\left(\phi_{\min} + \Delta\phi_{\min}, \phi_{\min}\right)$, as is shown in Fig.8d.

Thus, the method presented in section 3.1.1 can be used in monaxial rotation with uniform angular speed or accelerated angular speed.

3.1.2.2. Experimental Results of the Geometric-Moment-Based Algorithms for Monaxial Rotation Estimation

Experiment 4: To demonstrate the feasibility of the proposed method for monaxial rotation estimation, we designed an experimental platform to acquire real-world motion blurred images. This experimental platform generates monaxial rotations whose angular velocities are controllable by using a servo motor. A "white against black" mark is set on the head face of the output axis of the servo motor. The motion blurred images of the mark are captured by a BASLER camera whose axis is parallel to the rotational axis of the motor. The schematics of the experimental platform and part of the real experiment equipment are shown in Fig.7a and Fig.7b, respectively.

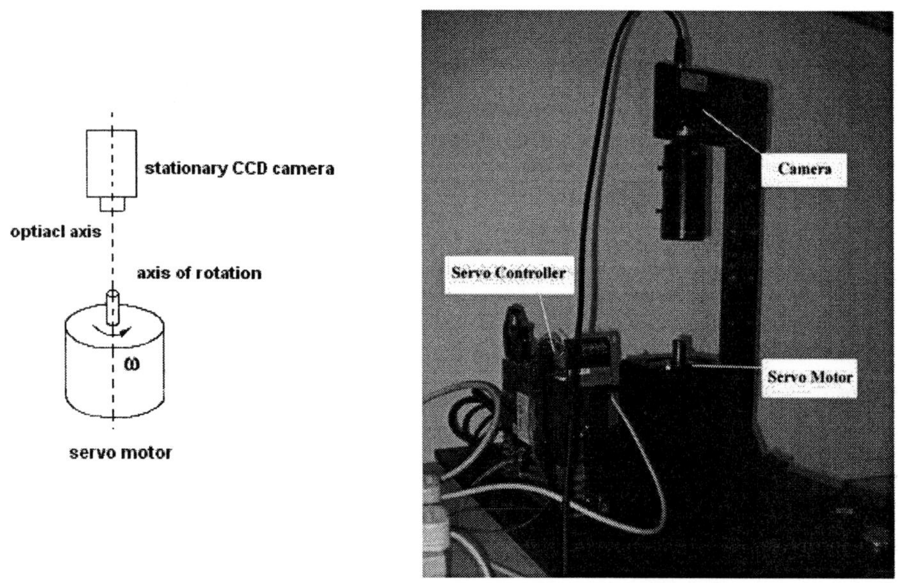

(a) (b)

Figure 7. (a) schematics of the experimental platform in Experiment 4; (b) part of the real experiment equipment.

In this experiment, we employ a motion blurred image and an unblurred reference image to calculate the angular velocities of rotations. The principle for estimating rotation from successive motion blurred images is similar to that using a motion blurred image and an unblurred reference image, so we give no unnecessary details here. We set the exposure time to be 42.66ms. The unblurred reference image of the mark is given in Fig.8a. Fig.8b and Fig.8c shows two motion blurred images, whose rotational speeds are 400r/min and 600 r/min, respectively. Fig.8d shows the projections of the "combined image" which is the summation of the unblurred image in Fig.8a and the motion blurred image in Fig.8b. The "combined image" for Fig.8a and Fig.8c is given in Fig.8e.

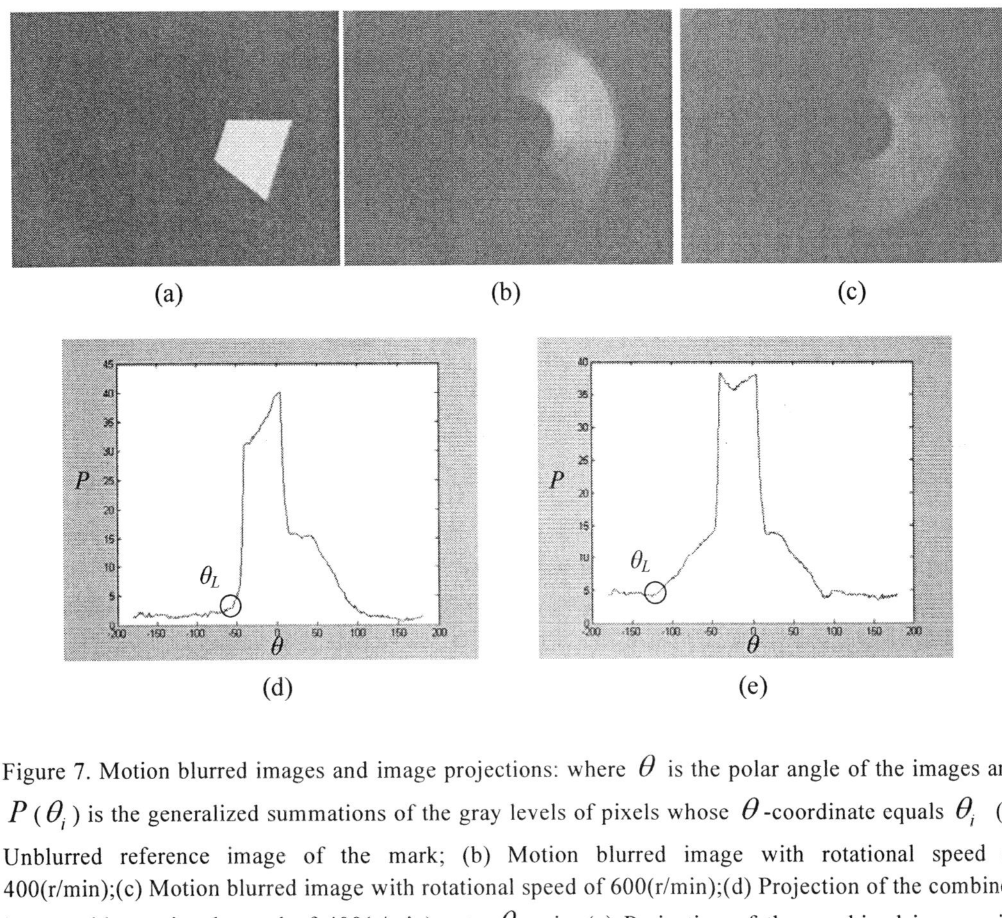

Figure 7. Motion blurred images and image projections: where θ is the polar angle of the images and $P(\theta_i)$ is the generalized summations of the gray levels of pixels whose θ-coordinate equals θ_i (a) Unblurred reference image of the mark; (b) Motion blurred image with rotational speed of 400(r/min);(c) Motion blurred image with rotational speed of 600(r/min);(d) Projection of the combined image with rotational speed of 400(r/min) onto-θ axis; (e) Projection of the combined image with rotational speed of 600(r/min) onto-θ axis.

Table 4. The experimental results for estimating monaxial rotations

Rotational Speed (in r/min)	Correct angular velocity ω^c (in degree/ms)	Measured angular velocity ω^e (in degree/ms)	Relative measured error of the angular speed%
400	2.4	2.357025	1.790612
600	3.6	3.478751	3.368039

The experimental results provided in Tab.4 shows that the proposed algorithm has good performance in monaxial rotation estimation.

3.2. Estimating Motion or Vibration Parameters from Motion Blur by Using Frequency Models

In frequency domain, translational motion blur usually forms certain stripe patterns in the spectrum of the motion blurred image. The spaces between two stripes are determined by the motion parameters. Thus, we can estimate a translational motion or vibration from a single frame of motion blurred image by analyzing the spaces between these stripes using image processing method. In this section, we discuss the theory and image processing method for some types of in-plane translational motions, such as uniform motion, high-frequency vibration, etc.

3.2.1. Theory for Estimating In-Plane Translational Uniform Motions Using Frequency Models

We use single frame to estimate in-plane translational uniform motion. Suppose the shutter opening time and exposure time for acquiring the motion blurred image are t_s and t_e respectively. Suppose the displacement vector $\mathbf{d}$ of the uniform image motion within the exposure time can be given by:

$$\mathbf{d} = \left(d_x, d_y\right)^T \tag{51}$$

where d_x and d_y denote the image displacements within the exposure time in x and y directions, respectively.

Then based on Eq. (6), a single blurred image $g(x, y)$ blurred by in-plane uniform motion is rewritten as:

$$g(x,y) = \frac{K_g}{K_f t_e} \int_{t_s}^{t_s+t_e} f\left(x - d_x t / t_e, y - d_y t / t_e\right) dt \tag{52}$$

After Fourier Transform:

$$\begin{aligned} G(u,v) &= \int_{-\infty}^{+\infty} \int_{-\infty}^{+\infty} g(x,y) \exp[-i2\pi(ux+vy)] dxdy \\ &= \int_{-\infty}^{+\infty} \int_{-\infty}^{+\infty} \left\{ \frac{K_g}{K_f t_e} \int_{t_s}^{t_s+t_e} f\left(x - d_x t / t_e, y - d_y t / t_e\right) dt \right\} \exp[-i2\pi(ux+vy)] dxdy \\ &= \frac{K_g}{K_f t_e} \int_{t_s}^{t_s+t_e} \int_{-\infty}^{+\infty} \int_{-\infty}^{+\infty} \left\{ f[x - d_x t / t_e, y - d_y t / t_e] \exp[-i2\pi(ux+vy)] dxdy \right\} dt \end{aligned} \tag{53}$$

Let $x = x + d_x t / t_e, y = y + d_y t / t_e$, the Equation (53) can be written as:

$$
\begin{aligned}
G(u,v) &= \frac{K_g}{K_f t_e} \int_{t_s}^{t_s+t_e} \int_{-\infty}^{+\infty} \int_{-\infty}^{+\infty} f(x,y) \exp\{-i2\pi[u(x + d_x t / t_e) + v(y + d_y t / t_e)\} dxdydt \\
&= \frac{K_g}{K_f t_e} F(u,v) \int_{t_s}^{t_s+t_e} \exp[-i2\pi(ud_x t / t_e + vd_y t / t_e)] dt
\end{aligned}
\tag{54}
$$

Here, we define $H(u,v)$ as:

$$H(u,v) = \frac{K_g}{K_f t_e} \int_{t_s}^{t_s+t_e} \exp[-i2\pi(ud_x t / t_e + vd_y t / t_e)] dt \tag{55}$$

Then, Eq. (54) can be simplified to be Eq. (55):

$$G(u,v) = F(u,v) H(u,v) \tag{56}$$

$H(u,v)$ is the point spread function (*PSF*) of motion or the Fourier Transform of the optical transfer function (OTF). Obviously, The $H(u,v)$ has many zeros, which can be written as:

$$\mu_i = \pi(ud_x + vd_y) t_e \tag{57}$$

If the angle between motion direction and horizontal axis is θ, let $d_x = \mathbf{d} \cos\theta$ and $d_y = \mathbf{d} \sin\theta$, then the zeros can be given by

$$\pi \mathbf{d}(u \cos\theta + v \sin\theta) = \mu_i \tag{58}$$

The standard value of zeroes μ_i can be deducted by their sinusoidal function. Based on this analysis, there is a series of darkness parallel lines in the spectrum that relate to the zeroes of $H(u,v)$. These lines and the direction angle θ can be extracted by applying line detection technique of image processing to the spectrum image. At the same time, these lines are vertical to the motion direction. Further, from Eq. (58), we get:

$$\mathbf{d} = \frac{\mu_i}{\pi(u \cos\theta + v \sin\theta)} \tag{59}$$

Therefore, the displacement can be calculated by $H(u,v)$'s zeroes. According to the relationship between the displacement and $H(u,v)$'s zeroes, the accuracy of the calculating result of the displacement relies on the precision of lines extraction of the spectrum image. The precision of line extraction is related to the numbers of parallel lines which can be detected. Further, mathematical expression of the $H(u,v)$'s zeroes indicates that the increasing of the displacement of the motion within the exposure time will cause more lines in the frequency domain, and the decreasing of the displacement of motion will reduce the number of these lines. Therefore, to choose proper exposure time may help to improve the accuracy of the motion estimation.

3.2.2. Theory for Estimating In-Plane Sinusoidal Translational Vibrations Using Frequency Models

In this section, we estimate high-frequency sinusoidal translational vibrations from one single frame of motion blurred images.

Suppose the displacement function $\mathbf{d}(t)$ of the sinusoidal image motion within time interval $[0,t_e]$ takes the following form:

$$\mathbf{d}(t)=\begin{bmatrix} d_x(t) \\ d_y(t) \end{bmatrix}=\begin{bmatrix} A\cos\theta\cos(2\pi f_V t) \\ A\sin\theta\cos(2\pi f_V t) \end{bmatrix},\ t\in[0,t_e] \tag{60}$$

where $d_x(t)$ and $d_y(t)$ are the displacements at time t of the vibrating image along *x-axis*, *y-axis* ; A and f_V are the amplitude and frequency of the image vibration; θ is the angle between motion direction and horizontal axis.

Based on Eq. (6), a single blurred image $g\ (x,y)$ blurred by in-plane translational vibration motion can rewritten as:

$$g(x,y)=\frac{K_g}{K_f t_e}\int_0^{t_e} f\left[x-A\cos\theta\cos(2\pi f_V t),y-A\sin\theta\cos(2\pi f_V t)\right]dt \tag{61}$$

After Fourier Transform:

$$G(u,v)=F(u,v)\frac{K_g}{K_f t_e}\int_0^{t_e}\exp\left\{-j2\pi\left[uA\cos\theta\cos(2\pi f_V t)+vA\sin\theta\cos(2\pi f_V t)\right]\right\}dt \tag{62}$$

Here, we focus on discussing high-frequency vibrations. If T_0 is the period of the vibration, t_e can be expressed as $t_e=(n+\varepsilon)T_0$, where ε is a very small value and n is a positive integral number. When $t_e >> T_0$ or ε is small enough, $t_e\approx nT_0$. Thus, the Eq. (62) can be written by:

$$G(u,v) = F(u,v)\frac{1}{nT_0}\int_0^{nT_0} \exp(-j2\pi f_V A\cos(2\pi f_V t))dt \tag{63}$$

The integration item on the right-side of Eq. (63) is a periodic function. Let:

$$J_0(2\pi f_V A) = \frac{1}{T_0}\int_0^{T_0} \exp(-j2\pi f_V A\cos(2\pi f_V t))dt \tag{64}$$

where J_0 is zero-order Bessel function of first kind. Then, Eq.(64) can be simplified to be:

$$G(u,v) = F(u,v)J_0(2\pi f_V A) \tag{65}$$

Due to the zeros of Bessel function, a series of parallel dark ripples l_i (i=1,2,3...) appear on the spectrum of motion blurred image. The positions of the dark ripples are corresponding to the zero values μ_i of Bessel function. And according to Equation (65), we can know the positions of zeros are exactly related to the amplitude and direction of harmonic vibration. Let $f_V = u\cos\theta + v\sin\theta$, then the zeros can be written by:

$$2\pi A(u\cos\theta + v\sin\theta) = \mu_i \tag{66}$$

The standard values of zeros μ_i are known by Bessel function. The linear strips l_i and direction angle θ can be extracted with line detection technique of image processing. Then amplitude A can be calculated from Equation (67):

$$A = \frac{\mu_i}{2\pi(u\cos\theta + v\sin\theta)} \tag{67}$$

Mathematically, Eq. (67) is valid only when exposure time t_e is longer than the period T_0, i.e., to determine the amplitude of high-frequency sinusoidal vibration from a single frame of motion blurred image, the blurred image must include at least one complete mode shape of vibration.

3.2.3. Image Processing Method

We use image processing method to extract the zeros' positions of the PSF of the motion blur in this chapter. The whole process to estimate the vibration amplitude by applying image processing method to the motion blurred image is described as follows:

(1)Transforming to frequency domain.

In order to transform the image from spatial domain to frequency domain, Fourier transformation is applied to images. Fig.9a is the Fourier transformation spectrum of the unblurred image shown in Fig. 2a, Fig.9b is spectrum of the blurred image with the velocity of 10 pixel/t_e shown in Fig. 2b, and Fig.9 c is spectrum of the blurred image with the velocity of 20 pixel/t_e shown in Fig. 2d. Fig.13a is the Fourier transformation spectrum of the static image shown in Fig.12a, Fig.13b is the Fourier transformation spectrum of the high-frequency sinusoidal vibration blurred image shown in Fig.12b, and Fig.13c is the Fourier transformation spectrum of the static image shown in Fig.12c. By comparing the spectrum shown in Fig.9a of the unblurred image with the spectrums given by Fig.9b and Fig.9c of the motion blurred image or comparing Fig.13a with Fig.13b and Fig.13c, the spectrums clearly have a series of dark parallel ripples, as is described in Section 3.2.1 and 3.2.2.

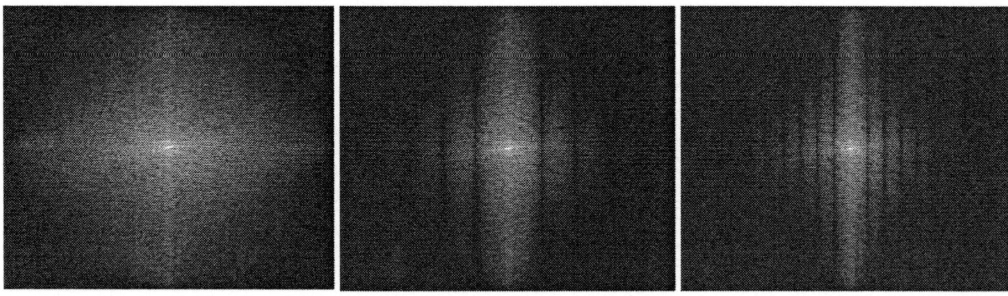

(a) (b) (c)

Figure 9. (a) The spectrum of the static image shown in Fig. 2(a). (b) The spectrum of the blurred image with the velocity of 10 pixel/t_e shown in Fig. 2(b). (c) The spectrum of the blurred image with the velocity of 20 pixel/t_e shown in Fig. 2(d).

(2)Lines segmentation.

Threshold is used to segment an image by setting all pixels whose intensity values are above a threshold to a foreground value and all the remaining pixels to a background value. Here, we assume that smaller image regions of ripples in the spectrum image are more likely to have similar intensity values. Local thresholds of these smaller image regions can be found by statistically examining the intensity values of the local neighborhood of each pixel. The method succeeds in the area surrounding these ripples because there are enough foreground and background pixels in the local neighborhood of each pixel, as are shown in Fig.10a.

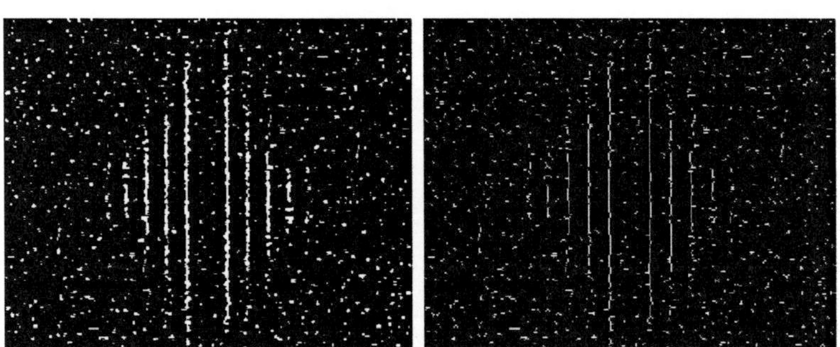

(a) (b)

Figure 10. (a) Lines segmentation result of Fig.9c. (b) Refining lines result.

(3)Refining lines.

We remove pixels on the boundaries of lines by refining lines to get the skeleton of ripples (shown in Fig.10b).

(4)Lines Detection.

We extract these parallel lines on spectrum using Radon transformation. After applying Radon Transform, 2D image with lines is transformed into a 1D curve, which gives peaks related to corresponding angle and lines. The position of each peak in the Radon transform curve is corresponding to a zero of Bessel function. We can transform the image twice by Radon transform: at first, we obtain the angle related to the vibrating direction by Radon transforming, as is shown in Fig.11a; then, we transform the image along the detected angle to obtain the position information of lines, as is shown in Fig.11b. It can be seen that several very bright spots are found in the Radon transform in Fig.11a, whose positions shown the parameters of the lines in the original image.

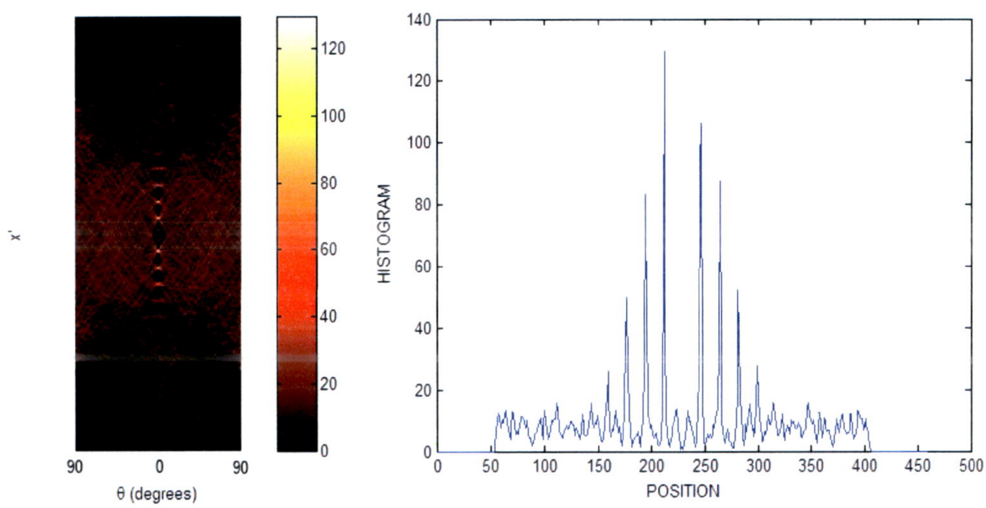

Figure 11. Radon transform.

(5)Lines' positions.

It's obviously that positions of ripples in spectrums are related to positions of the peaks in their Radon transforms. These peaks are the zero-crossings of the Randon transform curve's first order derivative and third order derivative. At the same time, the curve's second order derivative is negative. We extract these peaks of the Radon transform curve to detect the positions of ripples.

(6)Calculating results.

After positions of parallel lines are found, the vibration amplitude is obtained by Equation (57) or (67).

3.2.4. Experimental Results of the Spectrum-Analysis-Based Algorithms for Translational Motion Estimation

In this section, two experiments are designed to test the performance of the frequency-model-based motion estimation approaches. The first experiment is provided to verify the accuracy of the uniform motion estimation method. The second one is to validate the high-frequency vibration measurement approach.

3.2.4.1. Uniform Motion

Experiment 5: We applied the proposed method in Section 3.2 to the two frames of simulated motion blurred images, which is given in section 3.1(shown in Fig. 2b and Fig. 2d), to measure the performance of the proposed algorithm.

The experimental results of Experiment 5 are given in Tab.5. Tab.5 proves that this approach has good accuracy for estimating uniform motions.

Table 5. The experimental results for estimating uniform motion from a single frame of motion blurred image by frequency analysis*

Motion blurred images	v^c (in pixel/$t_e^{(g)}$)	v^e (in pixel/$t_e^{(g)}$)	ε^v (%)
Fig.2b	10	10.0571	0.571
Fig. 2d	20	20.1667	0.8335

3.2.4.2. Sinusoidal Motion

Experiment 6: We used the experimental platform referred in Section 3.1.1.4.3 to generate real sinusoidal translational vibrations and capture motion blurred images of the vibrating object for testing the performance of the presented method in vibration measurement. The exposure time is set to be 50ms. The sizes of the acquired images are 656×491. The proportion coefficient between the length unit (in pixel) of the images captured by this camera

Table 6. The experimental results for measuring sinusoidal vibrations by frequency analysis*

Motion blurred images	A^a (in mm)	θ^a (degree)	A^e (in mm)	θ^ε (degree)
Fig.12b	0.56mm	0	0.5598	0
Fig.12d	0.56mm	30	0.5617	30

* A^a and θ^a are the values measured by the accelerometer; A^e and θ^ε are the corresponding values measured by the proposed algorithm

(a) (b) (c)

Figure 12. (a) The static image. (b) The vibration blurred image: when acquiring it, the vibration amplitude and frequency of the object are 0.56mm and 100Hz, respectively. The exposure time is 50ms. (c) The vibration blurred image: when acquiring it, the vibration amplitude, the angle between vibrating direction and horizontal axis and frequency of the object are 0.56mm, 30 degree and 100Hz, respectively. The exposure time is 50ms.

and the physical length unit (in mm) is 0.027848mm/pixel. This experiment measures the parameters of a vibration by using a single frame of motion blurred image. The vibrational amplitude is 0.56mm shown in Fig.12b. And when capturing image shown in Fig.12c, the vibrational amplitude is 0.56mm and the angle between vibrating direction and horizontal axis is 30 degree.

The experimental results of Experiment 6 are given in Tab.6. The measured results from an accelerometer are also provided in Tab.6 for comparison. Tab.6 proves that the presented approach can measure the amplitude and direction of a high-frequency sinusoidal translational vibration effectively.

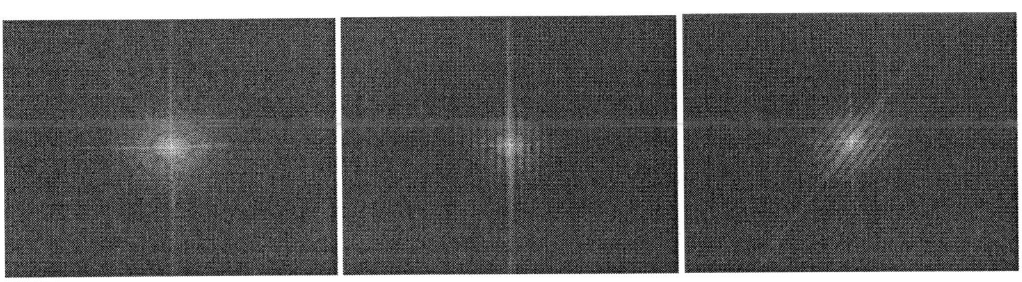

(a) (b) (c)

Figure 13. (a) The spectrum of the static image shown in Fig. 12a; (b) The spectrum of the blurred image shown in Fig. 12b; (c) The spectrum of the blurred image shown in Fig. 12c.

4. Conclusion

In this chapter, we get an inspiration from the "motion from blur" mechanism existing in the vision systems of some species, and present a novel group of approaches based on motion blurred images for estimating motion or vibration parameters in industrial applications. These approaches consider that image blur carries important motion information, and utilize this motion blur cue to estimate the parameters of motions or vibrations directly. The presented approaches involve both spatial-model-based method and frequency-model-based approach.

In spatial domain, we employ geometric moments to realize the extraction of motion blur information in blurred images. This chapter utilize some previously-derived and a newly-proved theorems, which reveal the relationship between the geometric moments of the translational motion blurred image and the displacement of the translational motion, to form nonlinear equations of the motion or vibration parameters. Then motions or vibrations can be estimated by solving these nonlinear equations.

In frequency domain, the spectrum of translational motion blurred image is multiplied by a Sinc function or a zero-order Bessel function for high-frequency vibration, which has zeros at positions related to the motion parameters. These zeros usually form stripes in the spectrum of the motion blurred image. We detect the positions of the zeros by employing image processing method on the stripe patterns to measure the parameters of motions or vibrations.

By representing the motion blurred images in a polar coordinate system, we can map a monaxial rotation into a 1D tranlstional motion. Then we can easily employ the translational motion estimation method to realize monaxial rotation measurement.

The proposed techniques can measure uniform motion, accelerated motion, and vibration from only one motion blurred image and an unblurred image for reference or from only two successive frames of motion blurred images, which is almost impossible for high-speed video or stroboscopic photography. Experiments with both synthetic and real motion blurred images are used to demonstrate the effectiveness of the presented methods in estimating motion and vibration.

Acknowledgement

It is gratefully acknowledged that the National Natural Science Foundation of China provided funds (NSFC 50375099) and Foundation of Bio-x DNA Cross-computer team (03DZ14025) for the financial support of this work.

References

Anderson C. H. and Van Essen, D. C., 1987, Shifter circuits: a computational strategy for dynamic aspects of visual processing. *Proc. Natl. Acad. Sci. USA* **84**, 6297–6301.

Arredondo, M.A., Lebart, K., Lane, D., 2004, Optical flow using textures, *Pattern Recognition Letters* **25**, 449–457

Barron, J. L., Fleet, D. J., Beauchemin, S. S., Burkitt, T.A., 1994, Performance of optical flow techniques. *Int. J. Compu. Vis.* **12**, 43-77

Ben-Ezra M., Nayar, S. K., 2004. *Motion-based motion deblurring. IEEE transactions on pattern analysis and machine intelligence* **26**, 689-698.

Borza, D.N., 2004. High-resolution time-average electronic holographyfor vibration measurement. *Optics and Lasers in Engineering* **41**, 515–527

Burns, D. J., Helbig, H. F., 1999. A system for automatic electrical and optical characterization of microelectromechanical devices. *Journal of microelectromechanical system* **8,** 473-482.

Burr, D. C., Ross, J., Morrone, M. C., 1986, Seeing objects in motion. *Proc. R. Soc. Lond. B Biol. Sci.* **227**, 249–265.

Chen, W. -G., Nandhakumar, N., Martin, W. N., 1996. Image motion estimation from motion smear-a new computational model. *IEEE Transactions on Pattern Analysis and Machine Intelligence* **18**, 412-424.

Davis, C.Q., Freeman, D.M., 1998. Using a light microscope to measure motions with nanometer accuracy. *Optical Engineering* **37**, 1299-1304.

Flusser, J., Suk, T., 1998. Degraded image analysis: An invariant approach. *IEEE Transactions on Pattern Analysis and Machine Intelligence* **20**, 590-603.

Gebert, G., Snyder, D., Lopez, J., Siddiqi, N., Evers, J., 2003, Optical Flow Angular Rate Determination, IEEE *International Conference on Image Processing* **1,** 949-952

Gregorio, R. D. and Parenti-Castelli, V., 2002, Mobility analysis of the 3-UPU parallel mechanism assembled for a pure translational motion. *Journal of Mechanical Design, Transactions of the ASME* **124**, 259-264.

Guan, B., Wang, S., Wang, G., 2005. A biologically inspired method for estimating 2D high-speed translational Motion. *Pattern Recognition Letters* **26**, 2450-2462

Hadar, O., Adar, Z., Cotter, A., et al., 1997. Restoration of image degraded by extreme mechanical vibrations. *Optics and Lasers in Engineering* **29**, 171-177.

Hafiane, A., Petitgrand, S., Gigan, O., et al., 2003. Study of sub-pixel image processing algorithms for MEMS in-plane vibration measurements by stroboscopic microscopy. *Proc. SPIE*, Vol. 5145, 169-179

Hemmers, W., Mermelstein, M.S., Freeman, D.M., 1999. Nanometer resolution of three-dimensional motions using video interference microscopy. *Proc. IEEE MEMS'99*, Orlando, Florida, pp.302-308.

Link, A., von Martens, H.-J., 1998. Amplitude and phase measurement of sinusoidal vibration in the nanometer range using laser interferometry. *Measurement* **24**, 55–67

Ming, Y., Haralick, R.M., Shapiro, L.G., 2002, Estimating optical flow using a global matching formulation and graduated optimization. *Proc. Int. Conf. on Image Process.*, 289–292.

Olaszek, P., 1999. Investigation of the dynamic characteristic of bridge structuresusing a computer vision method. *Measurement* **25**, 227–236.

Osten, W., Jüptner, W., Seebacher, S., Baumbach, T., 1999. The qualification of optical measurement techniques for the investigation of material parameters of microcomponents. *Proc. SPIE*, Vol. 3825, pp. 152-163.

Rothberg, S., Bell, J., 2004. On the application of laser vibrometry to translational and rotational vibration measurements on rotating shafts. *Measurement* **35**, 201–210.

Salbut, L., Kujawinska, M., 2001. The optical measurement station for complex testing of microelement. *Optics and Lasers in Engineering* **36**, 225-240.

Stern, A., Kopeika, N. S., 1997. Analytical method to calculate optical transfer functions for image motion and vibrations using moments. *Journal of the Optical Society of America A: Optics and Image Science, and Vision* **14**, 388-396.

Stern, A., Kempner, E., Shukrun, A., Kopeika, N. S., 2000. Restoration and resolution enhancement of a single image from a vibration-distorted image sequence. *Optical engineering* **39**, 2451-2457.

Tigges, M., Wittenberg, T., Mergell, P., et al., 1999. Imaging of vocal fold vibration by digital multi-plane kymography. *Computerized Medical Imaging and Graphics* **23**, 323–330.

Turner, K.L., Hatwell, P.G., MacDonald, N.C., 1999. Multidimensional MEMS motion characterization using laser vibrometry, *Proc. Transducers* **99**, pp. 1144-1147.

Warrant, E. J., 1999, Seeing better at night: life style, eye design and the optimum strategy of spatial and temporal summation. *Vis. Res.* **39**, 1611–1630

Ye, M., Haralick, R.M., Shapiro, L.G. 2003, Estimating piecewise-smooth optical flow with global matching and graduated optimization *IEEE Trans. Pattern Anal. Machine Intell.* **25**, 1625-1630

Yitzhaky, Y., Boshusha, G., Levy, Y., et al., 2000. Restoration of an image degraded by vibrations using only a single frame. *Optical Engineering* **39**, 2083-2091.

Zhang, Y. N., Wen, C.Y., Zhang, Y., 2000. Estimation of motion parameters from blurred images. *Pattern Recognition Letters* **21**, 425-433.

In: Pattern Recognition in Nanoscience...
Editor: Diane M. Perez, pp. 97-122
ISBN 1-60021-715-X

Chapter 4

THEORETICAL ANALYSIS OF LINEAR OPTICAL PROPERTIES OF SILICON CARBIDE NANOTUBES

***Vl.A. Margulis*[1]*, *E.E. Muryumin*[2] *and* *E.A. Gaiduk*[2]
[1]Department of Physics, N. P. Ogarev Mordovian State University, Saransk 430000, Russia
[2]Department of Chemistry, N. P. Ogarev Mordovian State University, Saransk 430000, Russia

Abstract

Recent development of nanoscience and nanotechnology has driven a tremendous research activity into tubular quasi-one-dimensional structures of nanometer size, such as carbon nanotubes, boron nitride nanotubes and many others, made of inorganic materials. More recently, in 2001 a novel kind of nanotubes was fabricated, namely, silicon carbide nanotubes (SiC–NTs), which are the subject of growing interest due to their potential applications in nanotechnology and their fundamental physical properties. From both perspectives, the study of optical response of SiC–NTs is of utmost importance. We report here a theoretical calculation of the linear optical coefficients (dielectric function, refractive index, reflectance and optical absorption coefficient) of single-walled SiC–NTs of different diameters, focusing on uniform ensembles of such tubes with a zigzag achiral structure. The calculation is carried out within a simple analytical model of the electronic structure of the SiC–NTs, based on an effective-mass approximation, and, at least to our knowledge, is the first for such nanotubes. The optical spectra we have calculated exhibit rich characteristic structures associated with interband electron transitions between pairs of van Hove singularities in the electronic density of states of the tubes. The collection of the results for the frequency-dependent linear optical coefficients, presented in this chapter, is a good test bed for the above-mentioned model of the electronic structure of the SiC–NTs and, at the same time, may be useful for developing physical insights into the optical properties of such tubes.

1. Introduction

Tubular forms of nanometer-sized quasi-one-dimensional (1D) materials have attracted a great deal of attention worldwide since the discovery of carbon nanotubes (CNTs) by a

*E-mail address: 612033@inbox.ru. (Corresponding author.)

NEC scientist Sumio Iijima in 1991 [1] (for a review of the basic concepts and the current status of the field, see [2–6]). Soon after this discovery, it has been realized that similar tubular structures could also be built with atoms of other elements, and afterwards a number of inorganic tubular forms of III–V compounds, such as BN [7–11], AlN [12], GaN [13], as well as nanotubes of metal dichalcogenides, such as WS_2, MoS_2 [14–16], and of metal oxides (e.g., VO_2 [17] and TiO_2 [18]), have been synthesized employing different techniques (for a recent review of the subject with an extensive list of references, see, e.g., [19]). Another interesting class of tubules motivated by graphite analogs, such as BC_2N and BC_3, are also available at present [20]. The properties of all the above-mentioned tubules are quite different from those of CNTs, suggesting numerious possible technological applications in various domains, such as catalysis, lubrication, electronic and photonic devices, etc.

Motivated by searching of promising candidates for a new generation of nanoscale electronic and optoelectronic devices, based on nanotubes, several experimental groups have recently succeeded in synthesizing silicon carbide nanotubes (SiC–NTs) [21–27], which are very suitable materials for applications in high-temperature harsh-environment nanoelectronics, nanosensors and nanoactuators, as well as for biotechnology applications. Theoretically, such tubes could be expected to be unstable, as initially predicted by Zhang et al. [28], due to different bonding characteristics that C and Si atoms exhibit in most stable configurations: the former are, as is well known, in an sp^2-hybridized state, while the latter are in an sp^3-hybridized one. However, both the generalized tight-binding molecular-dynamics and *ab initio* calculations [29, 30] have shown that SiC–NTs are stable when the concentration ratio of Si to C atoms is equal to 1, the energetically most favorable configuration consisting of alternating Si and C atoms with each Si atom having as nearest neighbors only C atoms and vice versa. The density-functional-theory calculations of the electronic structure of SiC–NTs corresponding to this configuration [31] have shown that all the SiC–NTs are semiconductors with the relatively wide band gap that depends on the helicity and diameter of the tubes. It is of general interest that, according to these calculations, chiral and armchair SiC–NTs are indirect-band-gap semiconductors, whereas SiC–NTs with a zigzag achiral structure have a direct band gap at the Γ point of the 1D Brillouin zone of the tubes. Thus, SiC–NTs are expected to have electronic properties quite different from the bulk SiC polytypes, which possess an indirect band gap [32], as well as from CNTs, which can be either metallic or semiconducting depending on their helicity and diameter [2–6]. The theoretical prediction by Zhao et al. [31] of a direct band gap in the zigzag SiC–NTs is especially important for the future development of semiconductor nano-optoelectronics, because it indicates that such tubes may exhibit strong photo- and electroluminescence. It is obvious that this can open up new opportunities for studying specific properties of SiC–NTs and designing novel nanodevices based on these nanotubes, for example, light emitting diodes. However, to our knowledge, the experimental observations corroborating the above-mentioned features of the electronic structure of SiC–NTs are still missing. Meanwhile, there appears to be little doubt that it can only be a matter of time before corresponding experiments are performed for these nanotubes. In particular, an investigation of the optical response of SiC–NTs might be of major importance for probing their electronic structure. Therefore, it seems to be worthwhile and timely enough to present theoretical information on the optical properties of SiC–NTs, which can be useful to guide experimental efforts and to understand more about the electronic structure of these tubes.

We report here a detailed theoretical analysis of the linear optical properties of single-walled SiC–NTs of different diameters, focusing on uniform ensembles of a special class of such tubes having a zigzag wrapping with 0° helicity. Our study is based on a single-particle approach that uses an effective-mass approximation to describe the electronic structure of the SiC–NTs. We present a convenient collection of the results for the frequency dependent linear optical coefficients (such as dielectric function, refractive index, reflectance and optical absorption coefficient) of several representative uniform ensembles of zigzag single-walled SiC nanotubes (SiC–SWNTs) of fairly large diameters. Our main result is that the optical spectra of these nanotubes show sharp characteristic features that are "split" due to band structure effects for each of the wrapped structures considered.

The rest of the chapter is organized as follows. In Sec. 2 we construct the model Hamiltonian that describes the electronic structure of the zigzag SiC–SWNTs in an effective-mass approximation. The following Sec. 3 provides analytical expressions for both the real and the imaginary parts of the linear optical susceptibility $\chi^{(1)}(\omega)$, which are used in Sec. 4 to calculate numerically the optical spectra of a number of uniform SiC–SWNT ensembles. Finally, we give a conclusion in Sec. 5.

2. Theoretical Model for the Electronic Structure of Single-Walled SiC Nanotubes

We begin by briefly reviewing the geometrical structure of SiC–NTs. Currently, such tubes are produced by using different techniques [21–27], including a substitutional reaction of Si atoms with multiwalled CNTs (used as templates). The template synthesis results in the new tubular structures – the multiwalled SiC nanotubes (SiC–MWNTs) with the concentration ratio of Si to C atoms equal to 1. The produced SiC–MWNTs have intertube space ranging from 3.8 to 4.5 Å, significantly lager than 3.4 Å normally observed in multiwalled CNTs [22]. The large spacings between the walls implies very weak intertube interactions and suggests that a SiC–MWNTs can be generally treated as an ensemble of diameter distributed SiC–SWNTs. Hence, the independent nanotube approximation seems to be reasonable for estimating the properties of the SiC–MWNTs, in particular, their electronic structure (at least for not too low electron energies). For this reason, a detailed knowledge of the properties of the SiC–SWNTs is of primary importance from the fundamental physics standpoint, independent of whether such tubes are synthesized nowadays. While these structures (individual SiC–SWNTs or well-ordered arrays of SiC–SWNTs) are yet to be experimentally realized, but, with the rapid progress of improved techniques in this area, it is reasonable to expect that experimental studies of the properties of such systems will not be far away. A first step in this direction has recently been made by Larkin and co-workers [23], who have managed to produce chemical-vapor-deposited polycrystalline (or amorphous) hollow, uniform SiC–SWNTs with an extremely large diameter of about 200 nm. One can hope that, with the improvement of experimental conditions and techniques, it will be possible to reduce the size of SiC–SWNTs down to $1 - 10$ nm, where their properties are expected to be substantially different from the bulk SiC due to the quantum confinement of electrons along the circumference direction of the tubes. Motivated by this

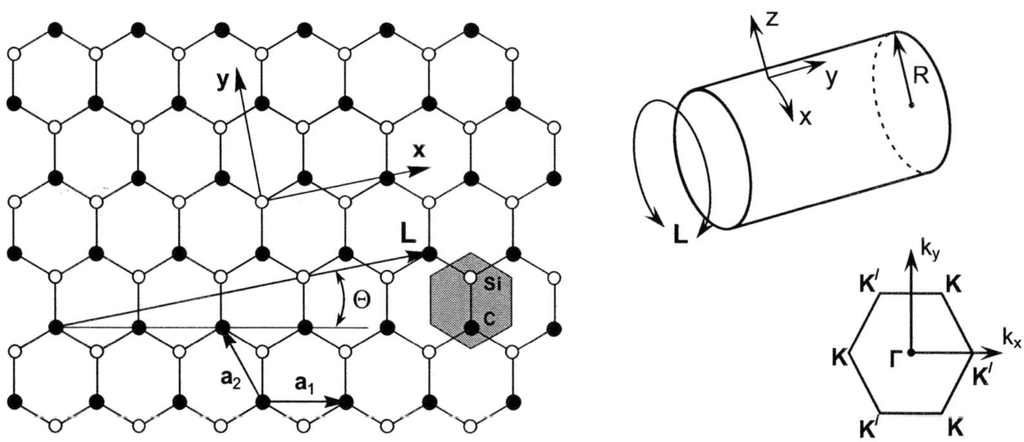

Figure 1. Lattice structure of a two-dimensional SiC sheet (left), geometrical structure of a single-walled SiC nanotube (right-top), and the first Brillouin zone of the SiC sheet (right-bottom). The shaded hexagon represents a unit cell of the SiC sheet. For the other notations, see the text.

expectation, we concentrate here on the discussion of the structure and properties of just such SiC–SWNTs.

The precise geometry of the SiC–SWNTs has not, however, been determined so far in experiment. For simplicity, we assume that the structure of the SiC–SWNTs is essentially graphitic with respect to the bonding of Si and C atoms and the orbital hybridization. This implies that, geometrically, a SiC–SWNT can be viewed as constructed by rolling up a strip of the graphene-like hexagonal SiC network, consisting of alternating Si and C atoms, into a cylinder. (The word "graphene" denotes a single carbon layer with the graphite structure.) Although in practice SiC–NTs are certainly not produced in that way, the idea to treat them as a wrapping SiC sheet is very appealing from the theoretical standpoint, since it allows their electronic band structure to be rationalized (see further below).

The wrapped structure of the nanotubes can be specified by the chiral vector

$$\mathbf{L} = l_1\mathbf{a}_1 + l_2\mathbf{a}_2, \tag{1}$$

where l_1 and l_2 are integers, $\mathbf{a}_1$ and $\mathbf{a}_2$ are the primitive lattice translation vectors of a two-dimensional honeycomb SiC sheet, making an angle of 120° (see Fig. 1). Choosing the Cartesian coordinate axes x and y in such a way that the corresponding unit vectors $\mathbf{e}_x$ and $\mathbf{e}_y$ are directed parallel and perpendicular to the vector $\mathbf{L}$, respectively, we have

$$\mathbf{a}_1 = a_0(1,0), \qquad \mathbf{a}_2 = a_0\left(-\frac{1}{2}, \frac{\sqrt{3}}{2}\right), \tag{2}$$

where the lattice constant $a_0 = |\mathbf{a}_1| = |\mathbf{a}_2|$ is equal to $\sqrt{3}$ multiplied by the Si–C bond length $d = 1.87$ Å [30].

Thus, one can specify a SiC–SWNT by a pair of indexes (l_1, l_2), in terms of which the nanotube radius R and the chiral angle Θ are expressed as follows:

$$R = \frac{a_0}{2\pi}\sqrt{l_1^2 + l_2^2 - l_1 l_2}, \tag{3}$$

$$\Theta = \tan^{-1}\left(\frac{\sqrt{3}l_2}{2l_1 - l_2}\right). \tag{4}$$

The two special classes of the (l_1, l_2) nanotubes are the so-called zigzag nanotubes with $(l, 0)$ indexes and the armchair tubes with (l, l) indexes. The nanotubes of both these classes are called achiral, whereas all the other tubes are called chiral. These is no experimental information available at present on the preferable structure of the SiC–SWNTs produced in the current synthesis processes [23]. The density-functional-theory calculations [30] show that both the above-mentioned achiral structures (zigzag and armchair) of the SiC–SWNTs are almost isoenergetic and, therefore, can be produced in experiments. However, their electronic band structure is predicted to be significantly different in that the armchair tubes are inderect-band-gap semiconductors, whereas the zigzag tubes have a direct band gap at the Γ point of the 1D Brillouin zone of the tubes [31]. It is this feature that makes, as remarked earlier, the zigzag SiC–SWNTs very attractive for a variety of optoelectronic applications. For this reason, we restrict ourselves to the consideration of just this class of achiral SiC–SWNTs.

The electronic structure of the zigzag SiC–SWNTs can be explored using a $\mathbf{k} \cdot \mathbf{p}$ two-band model within an effective-mass approximation as it has previously been done by Ajiki and Ando [33] for CNTs (for a review, see [34]) and by Král, Mele and Tománek [35,36] for BN nanotubes. This simplest model starts from the tight-binding Hamiltonian H for the π-electrons of the graphene-like SiC sheet, which is parametrized by a site diagonal potential Δ, which distinguishes the Si and C sites, and by the transfer integral $V_{pp\pi}$ between π-orbitals of nearest-neighboring Si and C atoms. These π-orbitals, aligned normal to the $x - y$ plane, form two π-bands (valence and conduction), the edges of which are located at the $\mathbf{K}$ and $\mathbf{K}'$ points of the two-dimensional Brillouin zone of the SiC sheet (see Fig. 1). We take the latter to be

$$\mathbf{K} = \frac{2\pi}{a_0}\left(\frac{1}{3}, \frac{1}{\sqrt{3}}\right), \qquad \mathbf{K}' = \frac{2\pi}{a_0}\left(\frac{2}{3}, 0\right). \tag{5}$$

Within the tight-binding approach, the periodic parts of the band-edge Bloch functions at the $\mathbf{K}$ and $\mathbf{K}'$ points can be expressed as follows:

$$U_{\mathbf{K}}^{(j)}(\mathbf{r}) = \frac{1}{\sqrt{N}}\sum_n \varphi(\mathbf{r} - \mathbf{R}_n - \mathbf{d}_j)\exp\{-i\mathbf{K}(\mathbf{r} - \mathbf{R}_n - \mathbf{d}_j)\}, \tag{6}$$

$$U_{\mathbf{K}'}^{(j)}(\mathbf{r}) = \frac{1}{\sqrt{N}}\sum_n \varphi(\mathbf{r} - \mathbf{R}_n - \mathbf{d}_j)\exp\{-i\mathbf{K}'(\mathbf{r} - \mathbf{R}_n - \mathbf{d}_j)\}, \tag{7}$$

where $j = 1(2)$ labels the sublattice occupied by Si(C) atoms, N is the total number of Si and C sites, $\varphi(\mathbf{r} - \mathbf{R}_n - \mathbf{d}_j)$ is the wave function of a normalized π-orbital for an atom located at $\mathbf{R}_n + \mathbf{d}_j$, $\mathbf{R}_n$ and $\mathbf{d}_j$ are the position vectors of the nth unit cell and the jth atom in it, respectively.

Since we are interested in the low-energy electronic states near the $\mathbf{K}$ and $\mathbf{K}'$ points, it seems to be appropriate to use the $\mathbf{k} \cdot \mathbf{p}$ approximation in order to determine those states. Introducing the two-dimensional wave vector $\mathbf{k}(k_x, k_y)$, measured relative to the $\mathbf{K}(\mathbf{K}')$

point, and expanding the tight-binding Hamiltonian H in small $\mathbf{k}$ around these points, we obtain the effective $\mathbf{k}\cdot\mathbf{p}$ Hamiltonian H_{eff}, which is given by the following 4×4 matrix:

$$H_{\text{eff}} = \begin{pmatrix} \Delta & \gamma\left(\hat{k}_x - i\hat{k}_y\right) & 0 & 0 \\ \gamma\left(\hat{k}_x + i\hat{k}_y\right) & -\Delta & 0 & 0 \\ 0 & 0 & \Delta & \gamma\left(\hat{k}_x + i\hat{k}_y\right) \\ 0 & 0 & \gamma\left(\hat{k}_x - i\hat{k}_y\right) & -\Delta \end{pmatrix}, \quad (8)$$

where $\gamma = \sqrt{3}|V_{pp\pi}|a_0/2$ is the $\mathbf{k}\cdot\mathbf{p}$ interaction parameter, $\hat{k}_x = -i\partial/\partial x$ and $\hat{k}_y = -i\partial/\partial y$ are the wave-number operators in the x and y direction, respectively.

The Schrödinger equation with the effective Hamiltonian (8) is written as

$$H_{\text{eff}}\mathbf{F}(\mathbf{r}) = E\mathbf{F}(\mathbf{r}), \quad (9)$$

where $\mathbf{F}(\mathbf{r})$ is the four-component eigenvector,

$$\mathbf{F}(\mathbf{r}) = \begin{pmatrix} \mathbf{F}_{\mathbf{K}}(\mathbf{r}) \\ \mathbf{F}_{\mathbf{K}'}(\mathbf{r}) \end{pmatrix} \quad (10)$$

with

$$\mathbf{F}_{\mathbf{K}}(\mathbf{r}) = \begin{pmatrix} \mathbf{F}^{(1)}_{\mathbf{K}}(\mathbf{r}) \\ \mathbf{F}^{(2)}_{\mathbf{K}}(\mathbf{r}) \end{pmatrix}, \qquad \mathbf{F}_{\mathbf{K}'}(\mathbf{r}) = \begin{pmatrix} \mathbf{F}^{(1)}_{\mathbf{K}'}(\mathbf{r}) \\ \mathbf{F}^{(2)}_{\mathbf{K}'}(\mathbf{r}) \end{pmatrix}. \quad (11)$$

The two sublattice components of $\mathbf{F}_{\mathbf{K}}(\mathbf{r})$ and $\mathbf{F}_{\mathbf{K}'}(\mathbf{r})$, labelled by the superscripts 1 and 2 in Eq. (11), are the slowly varying envelope functions which multiply the corresponding Bloch amplitudes $U^{(1,2)}_{\mathbf{K}}(\mathbf{r})$ and $U^{(1,2)}_{\mathbf{K}'}(\mathbf{r})$ at $\mathbf{K}$ and $\mathbf{K}'$ points to yield the electronic states at these points in the effective-mass approximation.

Use of Eqs. (8) and (10) in Eq. (9) gives the following two 2×2 $\mathbf{k}\cdot\mathbf{p}$ equations, which describe the electronic states of the π-bands near the $\mathbf{K}$ and $\mathbf{K}'$ points:

$$\left[\gamma\left(\sigma_x\hat{k}_x + \sigma_y\hat{k}_y\right) + \sigma_z\Delta\right]\mathbf{F}_{\mathbf{K}}(\mathbf{r}) = E^{(\mathbf{K})}\mathbf{F}_{\mathbf{K}}(\mathbf{r}), \quad (12)$$

$$\left[\gamma\left(\sigma_x\hat{k}_x - \sigma_y\hat{k}_y\right) + \sigma_z\Delta\right]\mathbf{F}_{\mathbf{K}'}(\mathbf{r}) = E^{(\mathbf{K}')}\mathbf{F}_{\mathbf{K}'}(\mathbf{r}), \quad (13)$$

where σ_x, σ_y and σ_z are the Pauli spin matrices,

$$\sigma_x = \begin{pmatrix} 0 & 1 \\ 1 & 0 \end{pmatrix}, \quad \sigma_y = \begin{pmatrix} 0 & -i \\ i & 0 \end{pmatrix}, \quad \sigma_z = \begin{pmatrix} 1 & 0 \\ 0 & -1 \end{pmatrix}. \quad (14)$$

The electronic structure of the zigzag SiC–SWNTs $(l,0)$ can be easily obtained by using Eqs. (12), (13) and imposing the periodic boundary conditions in the circumference direction (the x axis). For the envelope functions introduced above, these conditions can be written as [33, 34]

$$\mathbf{F}_{\mathbf{K}}(\mathbf{r}+\mathbf{L}) = \mathbf{F}_{\mathbf{K}}(\mathbf{r})\exp\left(-i\frac{2\pi\nu}{3}\right), \quad (15)$$

$$\mathbf{F}_{\mathbf{K}'}(\mathbf{r}+\mathbf{L}) = \mathbf{F}_{\mathbf{K}'}(\mathbf{r}) \exp\left(i\frac{2\pi\nu}{3}\right), \tag{16}$$

where $\nu(= 0, \pm 1)$ is the number uniquely determined by the relation $l = 3M + \nu$ with integer M.

Taking account of the two-component structure of both the functions $\mathbf{F}_{\mathbf{K}}(\mathbf{r})$ and $\mathbf{F}_{\mathbf{K}'}(\mathbf{r})$, the following *Ansatz* can be used as a solution for Eqs. (12) and (13):

$$\mathbf{F}_{\mathbf{K}(\mathbf{K}')}(\mathbf{r}) = \begin{pmatrix} C^{(1)}_{\mathbf{K}(\mathbf{K}')}(\mathbf{k}) \\ C^{(2)}_{\mathbf{K}(\mathbf{K}')}(\mathbf{k}) \end{pmatrix} \exp(i\mathbf{k}\mathbf{r}), \tag{17}$$

where $C^{(1)}_{\mathbf{K}(\mathbf{K}')}(\mathbf{k})$ and $C^{(2)}_{\mathbf{K}(\mathbf{K}')}(\mathbf{k})$ are functions of $\mathbf{k}$ to be determined. Now that we impose the periodic boundary conditions (15) and (16), the x-component of the electron wave vector $\mathbf{k}$ is quantized to the allowed discrete values

$$\kappa_{\nu m} = \frac{m}{R} - \frac{\nu}{3R}, \qquad m = 0, \pm 1, \pm 2, \ldots, \pm(l-1), \tag{18}$$

whereas k_y (further labelled simply k) remains continuous for an infinitely long nanotube and takes values within the one-dimensional Brillouin zone with a center at $k = 0$ (the Γ point) – namely, $k \in [-k_{\mathrm{BZ}}, k_{\mathrm{BZ}}]$, where $k_{\mathrm{BZ}} = \pi/a_0\sqrt{3}$ for a zigzag nanotube.

Use of the *Ansatz* (17) allows Eqs. (12) and (13) to be readily solved to yield the energy bands and the corresponding electronic eigenstates around the $\mathbf{K}$ and $\mathbf{K}'$ points. Near the $\mathbf{K}$ point, we obtain the π-electron-energy dispersion for the valence $(s = v)$ and conduction $(s = c)$ bands in the form

$$E_{s\mathbf{k}} = E_{s\nu mk} = \pm\sqrt{\Delta^2 + \gamma^2\left(\kappa^2_{\nu m} + k^2\right)}, \tag{19}$$

where the upper (lower) sign corresponds to the $c(v)$ band. Hereafter, for simplicity of the natation we omit the valley index $\mathbf{K}$.

The envelop functions for the two above-mentioned energy bands associated with the $\mathbf{K}$ point are given by

$$\mathbf{F}_{c\mathbf{k}}(\mathbf{r}) = \frac{1}{\sqrt{L}} \exp\left(i\kappa_{\nu m}x\right) \frac{1}{\sqrt{A}} \exp\left(iky\right) \begin{pmatrix} C^{(1)}_{c\mathbf{k}} \\ C^{(2)}_{c\mathbf{k}} \end{pmatrix}, \tag{20}$$

$$\mathbf{F}_{v\mathbf{k}}(\mathbf{r}) = \frac{1}{\sqrt{L}} \exp\left(i\kappa_{\nu m}x\right) \frac{1}{\sqrt{A}} \exp\left(iky\right) \begin{pmatrix} C^{(1)}_{v\mathbf{k}} \\ C^{(2)}_{v\mathbf{k}} \end{pmatrix}, \tag{21}$$

where A is the length of the nanotube, which is assumed to be much larger than L,

$$\begin{pmatrix} C^{(1)}_{c\mathbf{k}} \\ C^{(2)}_{c\mathbf{k}} \end{pmatrix} = \frac{1}{\sqrt{2}} \begin{pmatrix} v_{\mathbf{k}} \exp\left(-i\phi_{\mathbf{k}}/2\right) \\ u_{\mathbf{k}} \exp\left(i\phi_{\mathbf{k}}/2\right) \end{pmatrix} \exp\left(i\theta_{c\mathbf{k}}\right), \tag{22}$$

$$\begin{pmatrix} C^{(1)}_{v\mathbf{k}} \\ C^{(2)}_{v\mathbf{k}} \end{pmatrix} = \frac{1}{\sqrt{2}} \begin{pmatrix} u_{\mathbf{k}} \exp\left(-i\phi_{\mathbf{k}}/2\right) \\ -v_{\mathbf{k}} \exp\left(i\phi_{\mathbf{k}}/2\right) \end{pmatrix} \exp\left(i\theta_{v\mathbf{k}}\right), \tag{23}$$

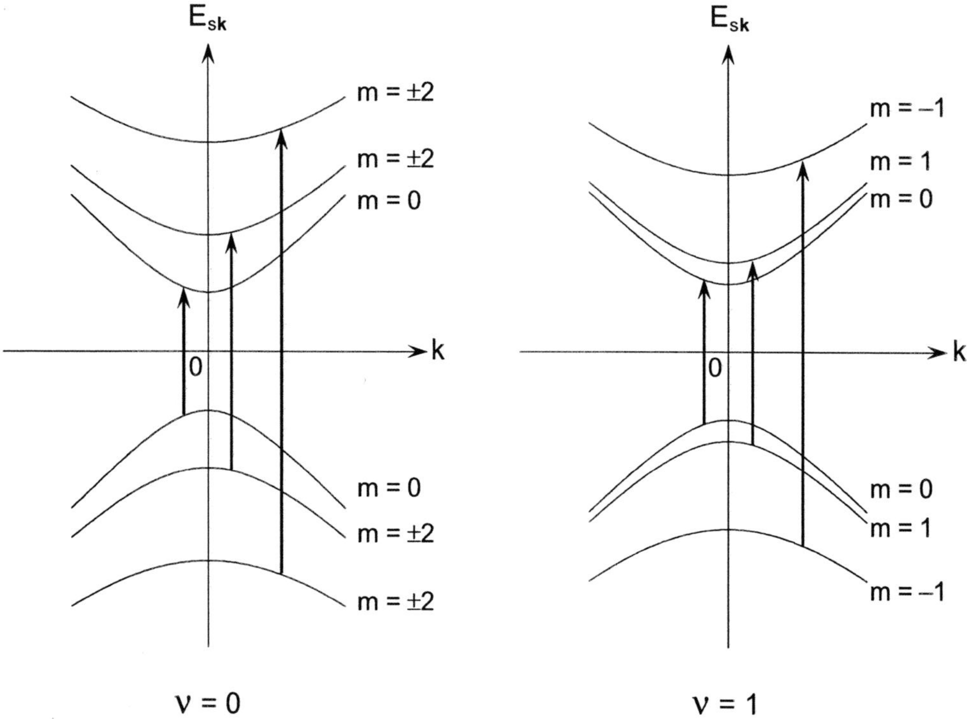

Figure 2. Schematic illustration of the energy subbands of a SiC–SWNT $(l, 0)$, which are obtained in the effective-mass approximation for $\nu = 0$ (left) and $\nu = 1$ (right). Vertical arrows indicate the optical transitions between the valence- and conduction-energy subbands with the same index m, electrically dipole-allowed for the parallel polarization of the incident light.

with

$$u_{\mathbf{k}} = \sqrt{\frac{E_{c\mathbf{k}} - \Delta}{E_{c\mathbf{k}}}}, \qquad v_{\mathbf{k}} = \sqrt{\frac{E_{c\mathbf{k}} + \Delta}{E_{c\mathbf{k}}}}, \tag{24}$$

$$\phi_{\mathbf{k}} = \tan^{-1}\left(\frac{k}{\kappa_{\nu m}}\right), \tag{25}$$

and there is also an arbitrary phase factor $\exp\left(i\theta_{s\mathbf{k}}\right)$ related to gauge invariance.

The energy bands associated with the $\mathbf{K}'$ point are obtained by the replacement $\nu \to -\nu$ in Eq. (19). Note also that this equation, as well as Eqs. (22) and (23) for the amplitudes of the envelope functions, are essentially the same as those obtained by Král, Mele and Tománek in their perspective paper [35] on BN nanotubes.

Thus, the band structure of the zigzag SiC–SWNTs, as derived from that of the SiC sheet using the simple zone-folding scheme [37–39] and the effective-mass approximation, represents a set of energy subbands specified by the angular-momentum quantum number m. Figure 2 shows schematically the energy dispersions of the several subbands in the vicinity of the Γ point of the 1D Brillouin zone of the nanotubes for $\nu = 0$ and $\nu = 1$. Note that at $\nu = 0$ all the subbands, except those with $m = 0$, are double degenerate in m, whereas at $\nu = \pm 1$ the degeneracy is lifted.

It is also interesting to note that, unlike the zigzag single-walled CNTs $(l, 0)$, which are metallic for $\nu = 0$ and semiconducting for $\nu = \pm 1$, the zigzag SiC–SWBTs are always semiconducting, independent of the value of ν. The latter affects only the width of the band gap E_g at the Γ point of the nanotube 1D Brillouin zone, onto which the $\mathbf{K}$ and $\mathbf{K}'$ points of the hexagonal two-dimensional Brillouin zone of the SiC sheet (Fig. 1) are mapped within the zone- folding method. Indeed, from Eq. (19) together with Eq. (18) it follows that

$$E_g = \begin{cases} 2\Delta & \text{for } \nu = 0, \\ 2\left(\Delta^2 + \dfrac{\gamma}{3R}\right)^{1/2} & \text{for } \nu = \pm 1. \end{cases} \tag{26}$$

Hence, for the zigzag SiC–SWNTs with $\nu = 0$ the band gap does not depend on the nanotube radius R at all, whereas in case $\nu = \pm 1$ it decreases with increasing R (but only very little). This is in sharp contrast to the results of the density-functional-theory calculations performed by Zhao et al. [31], who predicted that the band gap of SiC–NTs increases with increasing R. The discrepancy is puzzling and deserves thorough, investigation. One cannot exclude that the behavior of the band-gap energy E_g predicted by Zhao et al. [31] is an artifact of the computational method used in that paper. In this connection it is worthwhile to remind the reader that density-functional theory does not allow the band gap of semiconductors and dielectrics to be determined as accurately as the dispersion of the energy bands. For example, performing density-functional calculations Rurali [40] has obtained $E_g = 1.58$ eV for the bulk 3C–SiC against the experimental value of about 2.2 eV. On the other hand, it should be noted that the electronic band structure of the zigzag SiC–SWNTs presented above is based on the effective Hamiltonian (8) derived from the tight-binding Hamiltonian taking account of only π-atomic orbitals and as such neglects the $\sigma - \pi$ hybridization effects, which can turn out to be important for the nanotubes due to their finite curvature. Besides this, the misalignment of the π-orbitals at the nearest-neighbor Si and C atoms of the nanotubes is left out of account in our model. The above-mentioned effects are negligibly small for nanotube of somewhat large diameter (more than, say, 1 nm), but for the smallest nanotubes strong curvature-induced effects may significantly renormalize the band structure parameters of the tubes. However, in the forthcoming discussion we are not concerned with such thin tubes.

3. Linear Optical Susceptibility of Single-Walled SiC Nanotubes

We now turn to the calculation of the linear optical response of the zigzag SiC–SWNTs. We consider a uniform ensemble of such tubes that have their axes parallel to each other – as shown in Fig. 3. Let us designate by σ the surface density of the tubes in the plane perpendicular to their axes,

$$\sigma = AN_0/V, \tag{27}$$

where V is the normalized volume of the system composing N_0 identical tubes. Within the independent nanotube approximation, which we adopt in this work, the full response of the system to the external AC electromagnetic field can be obtained by the summation of the response of the individual tubes, contained in it. The macroscopic quantity that describes the response is the second rank tensor of the linear optical susceptibility $\chi^{(1)}_{\alpha\beta}(\omega)$, which in the case of an incident time-varying electric field $\mathbf{E}(t) = \mathbf{E}(\omega)\exp(-i\omega t)$ of optical frequency ω is determined by the following phenomenological relation:

$$P_\alpha(\omega) = \chi^{(1)}_{\alpha\beta}(\omega)E_\beta(\omega), \qquad \alpha, \beta = x, y, z, \tag{28}$$

where P_α is the α component of an induced polarization vector $\mathbf{P}$ inside the sample.

Within the microscopic theory, the interaction Hamiltonian of the electron in a nanotube with the electromagnetic perturbation can be written, in the dipole approximation, as

$$H_{\text{int}} = -e\mathbf{r}\mathbf{E}(t), \tag{29}$$

where $-e$ is the electron charge and $\mathbf{r}$ is the operator of the electron coordinate. From the band-theory formalism [41,42] it is well known that in crystal momentum representation $\mathbf{r}$ is decomposed into two parts as follows:

$$\mathbf{r} = i\frac{\partial}{\partial \mathbf{k}} + i\boldsymbol{\Omega}, \tag{30}$$

where the operator $\boldsymbol{\Omega}$ is determined by its matrix elements which are diagonal only in the wavevector $\mathbf{k}$:

$$\boldsymbol{\Omega}_{ss'}(\mathbf{k}) = \int U^*_{s\mathbf{k}}(\mathbf{r})\frac{\partial}{\partial \mathbf{k}}U_{s'\mathbf{k}}(\mathbf{r})\,d\mathbf{r}. \tag{31}$$

Here $U_{s\mathbf{k}}(\mathbf{r})$ is the periodic part of the Bloch wave function for the electronic states in band s with wavevector of $\mathbf{k}$. Thus, it is this part of the operator $\mathbf{r}$ that is relevant to the optical response associated with interband transitions.

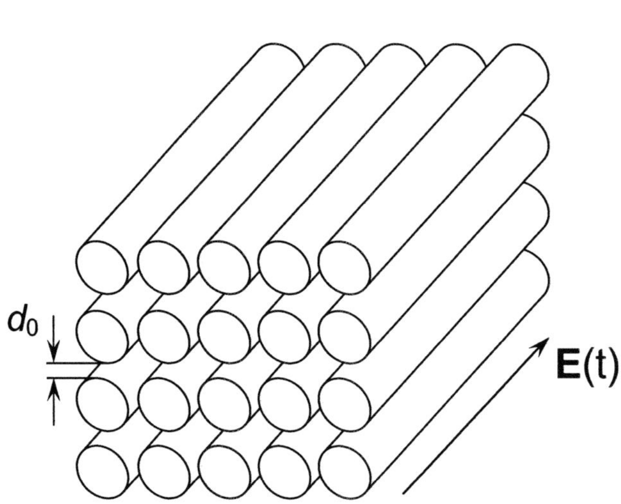

Figure 3. Schematic illustration of an array of identical and aligned SiC–SWNTs of radius R, arranged into a two-dimensional quadratic-lattice structure with the lattice constant $2R+d_0$. The arrow in the right indicates the direction of the time-oscillating electric field $\mathbf{E}(t)$ of a p-polarized monochromatic light beam incident normally onto this array.

Because of depolarization effects in nanotubes [43, 44], a strong optical response is expected to be observed only for light polarized along the tube axes. Therefore, in what follows we consider only this parallel polarization. Since only the one component of the susceptibility tensor $\chi^{(1)}_{\alpha\beta}(\omega)$ is then relevant – namely, $\chi^{(1)}_{yy}(\omega)$, in the following the subscript y will be dropped from the notation.

The selection rules for the optical transitions related to the electromagnetic radiation, which is linearly polarized along the tube axes, are $\Delta m = 0$ and $\Delta k = 0$ – i.e., only the vertical interband transitions, which couple the pairs of the electronic states with the same m, are electrically dipole-allowed for each k (see Fig. 2). These selection rules follow directly from the conversation law of the quantized angular momentum m, as well as from the fact that the photon momentum is extremely small on the microscopic scale.

We begin the calculation of the optical response by writting the general expression for the susceptibility $\chi^{(1)}(\omega)$. Such an expression was obtained in the past by a number of authors using different theoretical approaches. For our purpose it is convenient to use the expression for $\chi^{(1)}(\omega)$ derived by Genkin and Mednis [45], who developed a theory of the optical response (linear and nonlinear) of a two-band semiconductor system, based on the perturbation-theory formalism in the crystal-momentum representation. The calculation method suggested by Genkin and Mednis [45] is entirely within the framework of one-electron theory, i.e. the exciton effects and the local-field corrections were neglected. As has been shown in our previous papers [46, 47] on the optical properties of single-walled CNTs, the linear optical response of inhomogeneous ensembles of such tubes, calculated within the above-mentioned approach, is in good agreement with experiment. Therefore, there is good reason to believe that the independent-particle approximation could equally as well be applied to SiC–SWNTs. In any case, one-electron results can provide an invaluable reference point that allows a simple physical picture of the optical response to be established. Then by proceeding quite generally, the expression for $\chi^{(1)}(\omega)$ may be written as follows:

$$\chi^{(1)}(\omega) = -\frac{e^2}{V\hbar}\sum_{\mathbf{k}}\sum_{s}\sum_{s'\neq s} f_0(E_{s\mathbf{k}})\left[\frac{\Omega_{ss'}(\mathbf{k})\Omega_{s's}(\mathbf{k})}{\omega_{s's}(\mathbf{k})+\omega+i\Gamma}+\frac{\Omega_{ss'}(\mathbf{k})\Omega_{s's}(\mathbf{k})}{\omega_{s's}(\mathbf{k})-\omega-i\Gamma}\right], \quad (32)$$

where $f_0(E_{s\mathbf{k}})$ is the Fermi-Dirac distribution function, $\hbar\omega_{s's}(\mathbf{k}) = E_{s'\mathbf{k}} - E_{s\mathbf{k}}$ is the energy distance between the bands s and s' at fixed value of $\mathbf{k}$, and Γ is a phenomenological broadening factor regularizing the resonant divergencies, which is assumed to be independent of $\mathbf{k}$.

Turning now to the undefected zigzag SiC–SWNTs we are concerned with here, it should be noticed that, theoretically, their band-gap energy E_g was found to be rather large (of the order of $2-3$ eV [30,31,48]). Thus, E_g is much larger than the thermal energy $k_{\mathrm{B}}T$ (k_{B} is the Boltzmann constant) even at room temperature ($T_{\mathrm{room}} \simeq 26$ meV). Therefore, we assume that in such tubes the v-band is entirely filled, whereas the c-band is fully empty. This implies that it is legitimate to take $f_0(E_{s\mathbf{k}})$ in Eq. (32) to be 1 or 0 for the v-band and the c-band, respectively. In this case, we then get

$$\chi^{(1)}(\omega) = \frac{e^2}{V\hbar}\sum_{\mathbf{k}}|\Omega_{cv}(\mathbf{k})|^2\left[\frac{1}{\omega_{cv}(\mathbf{k})-\omega-i\Gamma}+\frac{1}{\omega_{cv}(\mathbf{k})+\omega+i\Gamma}\right]. \quad (33)$$

To calculate the matrix element $\Omega_{cv}(\mathbf{k})$ for the dipole interband transition, entering Eq. (33), we need to know the explicit expressions for the Bloch modulation amplitudes $U_{c\mathbf{k}}(\mathbf{r})$ and $U_{v\mathbf{k}}(\mathbf{r})$ for the two bands under consideration. Within the effective-mass approximation we have used above, both the amplitudes can approximately be expressed in terms of the periodic parts of the band-edge Bloch functions as follows:

$$U_{s\mathbf{k}}(\mathbf{r}) = \sum_{j=1,2} C_{s\mathbf{k}}^{(j)} U_{\mathbf{K}}^{(j)}(\mathbf{r}), \tag{34}$$

where $U_{\mathbf{K}}^{(j)}(\mathbf{r})$ is defined by Eq. (6) with N equal to

$$N = 2AL \Bigg/ \left(\frac{\sqrt{3}}{2} a_0^2 \right) = \frac{4AL}{\sqrt{3}a_0^2}. \tag{35}$$

Here $\sqrt{3}a_0^2/2$ is the area of the unit cell of the two-dimensional SiC sheet, and the pre-factor 2 takes into account that there are two atoms in the unit cell (see Fig. 1).

Inserting Eq. (34) into Eq. (31) and carrying out the integration in Eq. (31) leads to

$$\begin{aligned} \Omega_{cv}(\mathbf{k}) &\equiv \Omega_{cv}(\kappa_{\nu m}, k) \\ &= \frac{1}{2} \frac{\gamma^2}{E_{c\mathbf{k}} \sqrt{E_{c\mathbf{k}}^2 - \Delta^2}} \left(\frac{\Delta}{E_{c\mathbf{k}}} k - i\kappa_{\nu m} \right) \exp\left\{ -i(\theta_{c\mathbf{k}} - \theta_{v\mathbf{k}}) \right\}. \end{aligned} \tag{36}$$

After a bit of manipulation, we get

$$|\Omega_{cv}(\mathbf{k})|^2 = \frac{\gamma^2 \left(\Delta^2 + \gamma^2 \kappa_{\nu m}^2 \right)}{4E_{c\mathbf{k}}^4}. \tag{37}$$

Substituting Eq. (37) into Eq. (33) and converting the sum over k into the integral over the same variable by using the standard prescription

$$\sum_k \to 2 \frac{A}{2\pi} \int_{-k_{\mathrm{BZ}}}^{k_{\mathrm{BZ}}} dk \tag{38}$$

(the pre-factor 2 is due to spin degeneracy), we obtain

$$\begin{aligned} \chi^{(1)}(\omega) &= \frac{e^2 \gamma \sigma}{2\pi \Delta^2} \sum_{m=-(l-1)}^{l-1} \alpha_{\nu m}^2 \int_{\alpha_{\nu m}}^{\beta_{\nu m}} \frac{1}{x^3 \sqrt{x^2 - \alpha_{\nu m}^2}} \\ &\times \left(\frac{1}{x - z - i\delta} + \frac{1}{x + z + i\delta} \right) dx, \end{aligned} \tag{39}$$

where we have introduced the dimensionless variable x and the dimensionless parameters z and δ as follows:

$$x = \frac{E_{c\mathbf{k}}}{\Delta}, \qquad z = \frac{\hbar\omega}{2\Delta}, \qquad \delta = \frac{\hbar\Gamma}{2\Delta}. \tag{40}$$

In addition, the following notations are used in Eq. (39):

$$\alpha_{\nu m} = \sqrt{1 + \frac{\gamma^2 \kappa_{\nu m}^2}{\Delta^2}}, \tag{41}$$

$$\beta_{\nu m} = \sqrt{\alpha_{\nu m}^2 + \frac{\gamma^2 k_{\rm BZ}^2}{\Delta^2}}, \tag{42}$$

The integral in Eq. (39) can be rationalized by passing to the new variable of integration

$$t = \sqrt{\frac{x - \alpha_{\nu m}}{x + \alpha_{\nu m}}}. \tag{43}$$

Since $\chi^{(1)}(\omega)$ in Eq. (39) is complex, it needs to be separated into real and imaginary parts. Their derivation is straightforward, and, finally, we obtain

$$\begin{aligned} {\rm Re}\chi^{(1)}(\omega) &= \frac{e^2\gamma\sigma}{\pi\Delta^2} \sum_{m=-(l-1)}^{l-1} \frac{1}{\alpha_{\nu m}} \int_0^{\mu_{\nu m}} \left[(1-t^2)/(1+t^2)\right]^3 \\ &\times \left[P_{\nu m}(\delta, z, t) + P_{\nu m}(\delta, -z, t)\right] dt, \end{aligned} \tag{44}$$

$$\begin{aligned} {\rm Im}\chi^{(1)}(\omega) &= \delta\frac{e^2\gamma\sigma}{\pi\Delta^2} \sum_{m=-(l-1)}^{l-1} \frac{1}{\alpha_{\nu m}} \int_0^{\mu_{\nu m}} \left[(1-t^2)^4/(1+t^2)^3\right] \\ &\times \left[Q_{\nu m}(\delta, z, t) - Q_{\nu m}(\delta, -z, t)\right] dt, \end{aligned} \tag{45}$$

where

$$\mu_{\nu m} = \sqrt{\frac{\beta_{\nu m} - \alpha_{\nu m}}{\beta_{\nu m} + \alpha_{\nu m}}}, \tag{46}$$

$$P_{\nu m}(\delta, z, t) = D_{\nu m}(z, t) Q_{\nu m}(\delta, z, t), \tag{47}$$

$$Q_{\nu m}(\delta, z, t) = \left[D_{\nu m}^2(z, t) + \delta^2(1-t^2)^2\right]^{-1}, \tag{48}$$

$$D_{\nu m}(z, t) = \alpha_{\nu m}(1+t^2) - z(1-t^2). \tag{49}$$

Although Eqs. (44) and (45) are rather cumbersome, they have important computational advantages in that they are readily programmed to enable numerical results to be obtained, and these are shown in the next section for a number of the so-called optical-response functions. The latter involve the real and imaginary parts of the complex refractive index $\mathcal{N}(\omega)$ given by

$$[\mathcal{N}(\omega)]^2 = [n(\omega) + i\kappa(\omega)]^2 = \varepsilon(\omega), \tag{50}$$

where the complex dielectric function

$$\varepsilon(\omega) = \varepsilon_1(\omega) + i\varepsilon_2(\omega) \tag{51}$$

is related to the susceptibility $\chi^{(1)}(\omega)$ as

$$\varepsilon(\omega) = 1 + 4\pi\chi^{(1)}(\omega), \tag{52}$$

and, hence, the real index of refraction $n(\omega)$ and the extinction coefficient $\kappa(\omega)$, usually called the linear optical constants, can be expressed in terms of $\varepsilon_1(\omega)$ and $\varepsilon_2(\omega)$ as follows:

$$n^2(\omega) = \frac{\varepsilon_1(\omega)}{2}\left\{\left[1+\left(\frac{\varepsilon_2(\omega)}{\varepsilon_1(\omega)}\right)^2\right]^{1/2}+1\right\}, \tag{53}$$

$$\kappa^2(\omega) = \frac{\varepsilon_1(\omega)}{2}\left\{\left[1+\left(\frac{\varepsilon_2(\omega)}{\varepsilon_1(\omega)}\right)^2\right]^{1/2}-1\right\}. \tag{54}$$

One more important quantity that contributes much to the detailed understanding of the optical properties of a medium is the reflectance $\mathcal{R}(\omega)$, which in the case of plane-electromagnetic-wave propagation along the z axis (a normal incidence) can be calculated using the formula

$$\mathcal{R}(\omega) = \frac{[n(\omega)-1]^2+\kappa^2(\omega)}{[n(\omega)+1]^2+\kappa^2(\omega)}. \tag{55}$$

The optical absorption coefficient $\mathcal{K}(\omega)$ is then defined by the equation

$$\mathcal{K}(\omega) = \frac{\omega\,\varepsilon_2(\omega)}{c\;n(\omega)} \tag{56}$$

(c is the speed of light in free space), implying that the intensity of the incident wave is attenuated exponentially according to the law $\exp\{-\mathcal{K}(\omega)z\}$, where z is the distance into the medium along the line of wave propagation.

It is worthwhile to note that in the long-wavelength limit ($\omega \rightarrow 0$), where the susceptibility $\chi^{(1)}(\omega)$ is purely real, one can obtain an explicit analytical expression for $\chi^{(1)}(0)$ in terms of elementary functions of the three parameters γ, Δ and R. the relevant formula, obtained from Eq. (39) at $z=0$ and $\delta=0$, is as follows:

$$\chi^{(1)}(0) = \frac{e^2\gamma\sigma}{3\pi\Delta^2}\sum_{m=-(l-1)}^{l-1}\mathcal{F}\left(\frac{\Delta}{\gamma k_{\mathrm{BZ}}},\frac{\kappa_{\nu m}}{k_{\mathrm{BZ}}}\right), \tag{57}$$

where the function $\mathcal{F}(a,b)$ has the form

$$\mathcal{F}(a,b) = \frac{a^2(2+3a^2+3b^2)}{(a^2+b^2)(a^2+b^2+1)^{3/2}}. \tag{58}$$

Since the zero-frequency susceptibility $\chi^{(1)}(0)$ appears in evaluating the index of refraction n_0 in the transparency region,

$$n_0 \equiv n(\omega=0) = \left[1+4\pi\chi^{(1)}(0)\right]^{1/2}, \tag{59}$$

the above-mentioned formula for $\chi^{(1)}(0)$ can be useful in order to make simple estimates and qualitative predictions regarding the optical response of SiC–SWNT samples over the frequency range corresponding to virtual interband π-electron transitions. Further below we will yet return to the discussion of this point.

4. Dispersion of the Linear Optical Coefficients: Numerical Results and Discussion

In this section we present the results of our numerical calculations of the above-mentioned optical coefficients characterizing the linear optical properties of uniform ensembles of the zigzag SiC–SWNTs. All the calculations have been carried out using the following values of the parameters entering our model: $V_{pp\pi} = -1.9$ eV, $\Delta = 1.2$ eV, and $\hbar\Gamma = 24$ meV. Although the order of magnitude of these parameters seems to be reasonable, it should be stressed that none of them is inferred from experiment and neither is it taken to be adjusted to some experiment. However, we believe that it is of minor importance at this stage, when there still are no experimental data available on the optical properties of SiC–NTs. While at the moment we do not strive towards an agreement with possible experiments, we would like to note that the above-mentioned parameters can be varied at will, if it were necessary to get the best fit to experimental data. On the other hand, it is hard to expect that the qualitative features of the optical spectra, presented further below, could be changed significantly on varying the values of the model parameters.

We start our numerical study by considering the static dielectric constant $\varepsilon_0 = 1 + 4\pi\chi^{(1)}(0)$ and the index of refraction $n_0 = \sqrt{\varepsilon_0}$ at zero frequency – i.e., in the transparency region. Using Eq. (57) we have calculated these constants for a number of uniform ensembles of the zigzag SiC–SWNTs $(l, 0)$ with $l = 7 - 25$. We present the results obtained in Fig. 4, where the values of ε_0 and n_0 are plotted versus the index l for three cases differing from one another only by the way in which the tubes are arranged in the ensembles. The first case corresponds to densely packed tubes without gaps between them. The surface concentration σ of aligned SiC–SWNTs in the plane perpendicular to their axes is then given by

$$\sigma = 1 \big/ \pi R^2 \, . \tag{60}$$

The two remaining cases correspond to arrays of the same nanotubes arranged in two-dimensional quadrate and triangular lattices, respectively, with the minimal distance between the tubes d_0 approximately equal to the distance of about 4 Å between adjacent walls in a SiC–MWNT. For a SiC–SWNT array with a square-lattice structure (see Fig. 3), the σ is given by

$$\sigma = 1 \big/ (2R + d_0)^2 \, , \tag{61}$$

whereas in the case of a triangular-lattice structure

$$\sigma = 2 \big/ \sqrt{3}\,(2R + d_0)^2 \, . \tag{62}$$

It is worthwhile to note that the results obtained for the two last cases are shown in Fig. 4 only for illustrative purposes, since we are not aware of any experimental data pointing to the possibility of the formation of such crystalline bundles of SiC–SWNTs.

As seen from Fig. 4, both ε_0 and n_0 decrease with an increase of l (or, which is one and the same, with increasing the nanotube radius R) independent of the type of the nanotube arrangement in the ensembles. Moreover, on the scale of Fig. 4, we see no significant difference between the ε_0 values, as well as between n_0 ones, for the quadratic- and triangular-lattice structures, both ε_0 and n_0 exhibiting almost linear variation with l. In contrast, the l

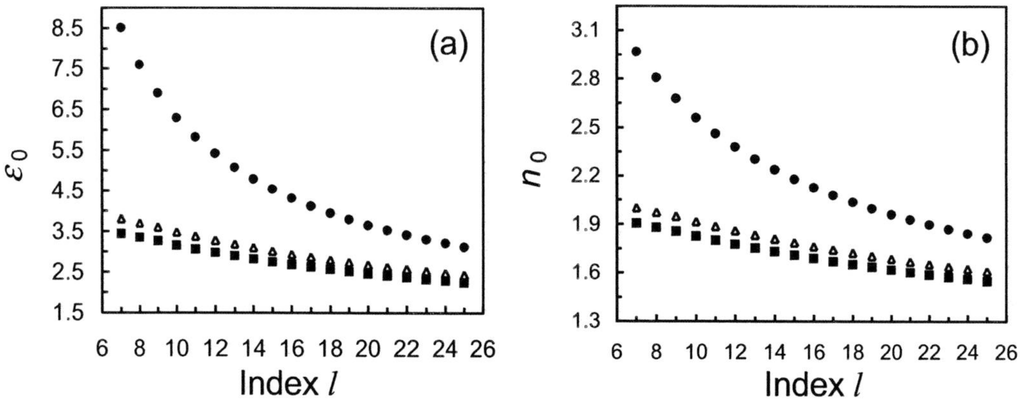

Figure 4. (a) Static dielectric constant ε_0 calculated for uniform ensembles of SiC–SWNTs $(l, 0)$ with l ranging from 7 to 25, and (b) refractive index n_0 at zero frequency for the same nanotube ensembles. Solid dots, squares and open triangles show the results obtained for the three types of SiC–SWNT ensembles specified by Eqs. (60), (61) and (62), respectively.

(or R)-dependence of ε_0 and n_0 in the case of the closely-packed structure of the tubes is essentially nonlinear, at least for small-diameter tubes. This effect arises from the presence of σ as a proportionality factor in the right-hand side of Eq. (57) for $\chi^{(1)}(0)$. Indeed, the function $\mathcal{F}(\Delta/\gamma k_{\mathrm{BZ}}, \kappa_{\nu m}/k_{\mathrm{BZ}})$ in that equation depends on R via the parameter $\kappa_{\nu m}$ [see Eq. (18)], but only weakly because of the specific form of the function $\mathcal{F}(a, b)$ in Eq. (58). As a result, the R-dependence of the zero-frequency susceptibility $\chi^{(1)}(0)$, as well of the ε_0 and the n_0, is dominated by the density σ, which is inversely proportional to the square of the nanotube radius in the case of the closely-packed structure of the nanotubes. However, a warning message may be in order in this connection. Presenting the results for the smallest nanotubes (with $l = 7 - 9$) in Fig. 4, we would like to mention that the consideration of the latter actually goes beyond the applicability of the theory developed in the preceding section because for such tubes a curvature effect is probably very important, and, hence, the zone-folding approximation used in that section may turn out to be insufficient.

It is noteworthy that the behavior of ε_0 and n_0 for an ensemble of closely-packed SiC–SWNTs, discussed above, is quite different from that predicted for single-walled CNT bundles of similar structure, where ε_0 and n_0 are shown to be independent of R [46, 47]. This directly reflects the difference between the electronic structure of these two types of nanotubes: contrary to the SiC–SWNTs, which are wide-gap semiconductors with an almost constant band gap independent of their diameters, the semiconducting single-walled CNTs $(l, 0)$ have a band gap, which is approximately inversely proportional to the nanotube diameter.

Finally, it is worthwhile to note that Fig. 4, making the specific prediction regarding the character of the dependence $n_0(l)$ for the three above-mentioned SiC-SWNT structures, suggests that at least the closely-packed structure and the two others considered may be distinguished in experiment by measuring the refractive index of the SiC–SWNT sample in the transparency region. A similar possibility of optical characterization of single-walled CNT samples has recently been reported by Tatsuura et al. [49].

We now turn to a discussion of the frequency dispersion of the linear optical coeffi-

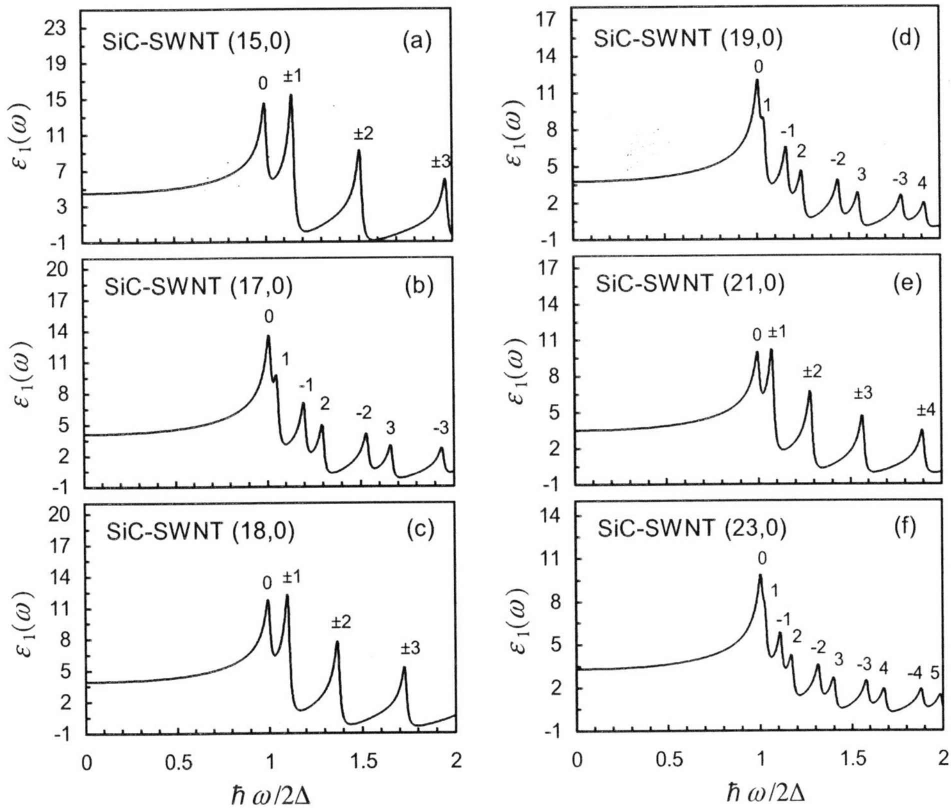

Figure 5. The real part of the dielectric function, $\varepsilon_1(\omega)$, calculated for six uniform SiC–SWNT ensembles consisting, respectively, of the zigzag tubes with chirality indices (15,0), (17,0), (18,0), (19,0), (21,0) and (23,0). The latter ones are indicated in the upper part of the panels. The numbers $0, \pm 1, \pm 2, \ldots$ on the top of the peaks correspond to our labelling convention for the interband electron transitions (for details, see the text).

cients, defined by Eqs. (51)–(56). In Figs. 5–9, the numerically calculated optical spectra $\varepsilon_1(\omega)$, $\varepsilon_2(\omega)$, $n(\omega)$, $\mathcal{R}(\omega)$ and $\mathcal{K}(\omega)$ are shown for six representative ensembles consisting, respectively, of the zigzag SiC–SWNTs (15,0), (17,0), (18,0), (19,0), (21,0) and (23,0). It should be noted that all the spectra have been calculated by using Eqs. (44), (45) and (60) and, thus, are formally relevant to ensembles of the closely-packed SiC–SWNTs. Yet the main qualitative features of the spectra do not depend on how the tubes are arranged in the ensembles, since within the independent-nanotube approximation the density σ enters Eqs. (44) and (45) only as a proportionality factor. Therefore, the spectra in Figs. 5–9 are representative ones not only for the closely-packed nanotube structure, but for the two other above-mentioned structures as well.

Figures 5–9 show the optical coefficients against the dimensionless photon energy $\hbar\omega/2\Delta$. Before we proceed to discuss these figures separately, we would like to mention that all the spectra plotted Figs. 5–9 are featureless below the threshold value $\hbar\omega/2\Delta = 1$ and demonstrate a progressive increase of the optical coefficients as $\hbar\omega/2\Delta$ approaches the threshold. In the region above the threshold, characteristic sharp peaks are observed in

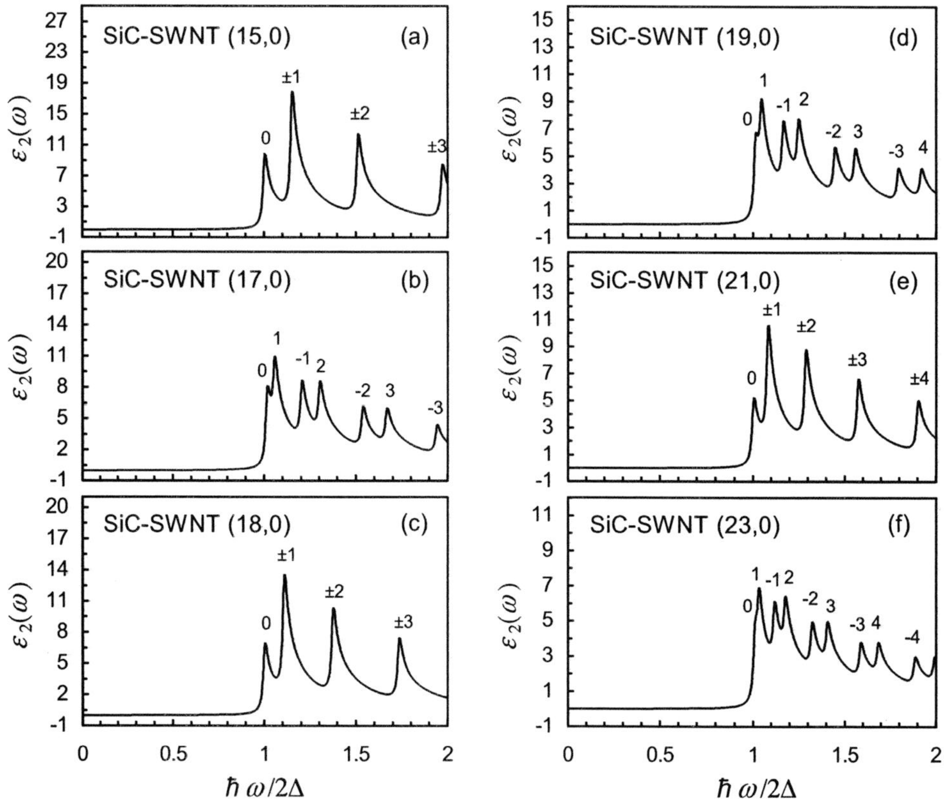

Figure 6. The imaginary part of the dielectric function, $\varepsilon_2(\omega)$, for the same SiC–SWNT ensembles, as in Fig. 5. Resonant interband electron transitions are indicated by the numbers on the top of the peaks, as in Fig. 5.

all the spectra, which can be assigned to electrically dipole-allowed vertical optical transitions between successive pairs of van Hove singularities in the one-dimensional electronic density of states of SiC–SWNTs. The mirror pairs of these singularities occur at the edges of the valence- and conduction-energy subbands with the same azimutal indices m. We will further use the abridged notation $v_m \to c_m$ for the corresponding interband electron transitions, labelling the corresponding peaks in Figs. 5–9 simply $m(= 0, \pm1, \pm2, \ldots)$. As has already been mentioned (see also Fig. 2), the energy subbands with $m \neq 0$ are doubly degenerate in m for SiC–SWNTs $(l, 0)$ with $l = 3M$ (further referred to as type-I nanotubes), while the degeneracy is lifted in the case of SiC–SWNTs $(l, 0)$ specified by the index $l = 3M + 1$ (further referred to as type-II nanotubes). As a consequence, the number of peaks observed in the optical spectra for the ensembles of the type-II nanotubes is larger than that for the ensembles of the type-I nanotubes. There is one more overall trend clearly seen in Figs. 5–9 – namely, for the thicker tubes, belonging to one and the same above-mentioned type of SiC–SWNTs, there are more peaks in the spectra, whereas the energy spacing between two neighboring peaks becomes smaller.

Let us now turn to Figs. 5 and 6, which illustrate the spectra of the real (ε_1) and the imaginary (ε_2) parts of the dielectric function $\varepsilon(\omega)$, respectively. Comparing the spectra for

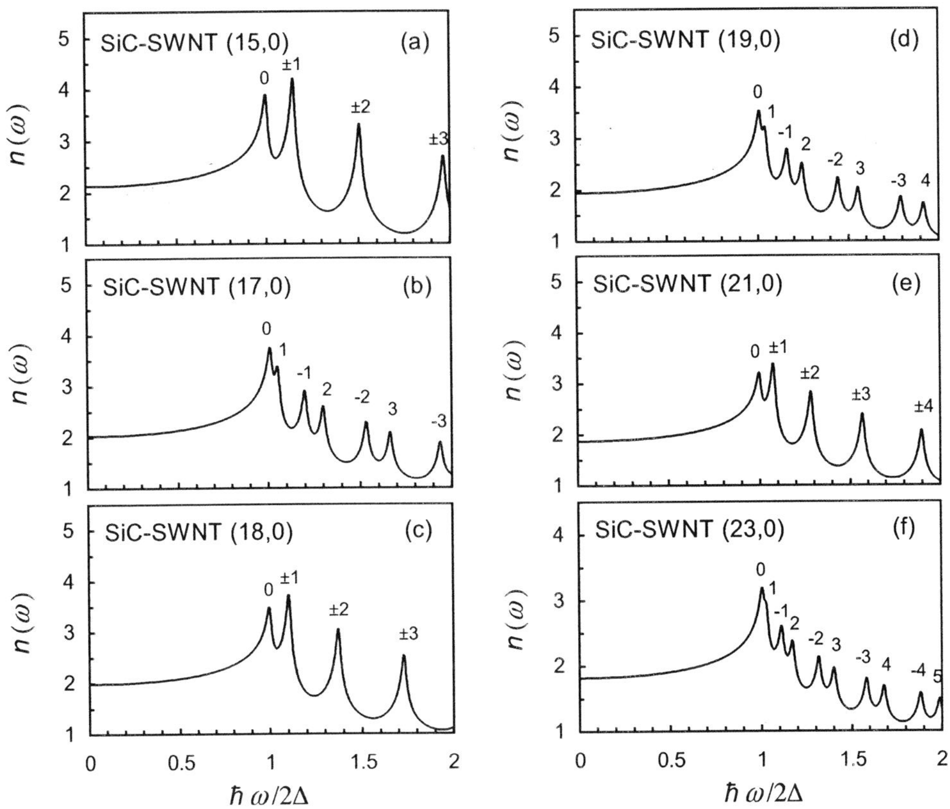

Figure 7. Calculated spectra of the refractive index $n(\omega)$ for the same SiC-SWNT ensembles, as in Fig. 5. Resonant interband electron transitions are indicated by the numbers on the top of the peaks, as in Fig. 5.

the type-I- and type-II-nanotube ensembles, we see that there are two important difference between them. First, as seen from Figs. 5a, 5c and 5e, the central, most prominent peak in the $\varepsilon_1(\omega)$ spectra of the type-I-nanotube ensembles actually consists of two subpeaks of almost equal hight, which correspond to $v_0 \rightarrow c_0$ and $v_{\pm 1} \rightarrow c_{\pm 1}$ transitions, respectively. In contrast, in the $\varepsilon_1(\omega)$ spectra of the type-II-nanotube ensembles (Figs. 5b, 5d and 5f), the central peak has no double-peak structure. Instead of this, in Fig. 5b we see a small cusp on the right-hand wing of the central peak, which is related to the $v_1 \rightarrow c_1$ transition. In going over to the larger diameter type-II nanotubes, the cusp is smeared out, transforming gradually to a narrow, hardly noteceable shoulder (Figs. 5d and 5f).

The second of the above-mentioned differences is that the $\varepsilon_1(\omega)$ spectra of the type-I-nanotube ensembles exhibit sharply asymmetrical, deep dip structures situated between all the nearest-neighboring peaks except the first two (Figs. 5a, 5c and 5e). In the right-hand neighborhood of the peaks the magnitude $\varepsilon_1(\omega)$ changes from large positive values to small positive (or negative) ones. Although such dip structures exist in the $\varepsilon_1(\omega)$ spectra of the type-II-nanotube ensembles as well (Figs. 5b, 5d and 5f), they are shallow rather than deep.

The above-mentioned differences in the $\varepsilon_1(\omega)$ spectra of the type-I- and type-II-nanotube ensembles have their counterparts in the corresponding $\varepsilon_2(\omega)$ spectra shown in

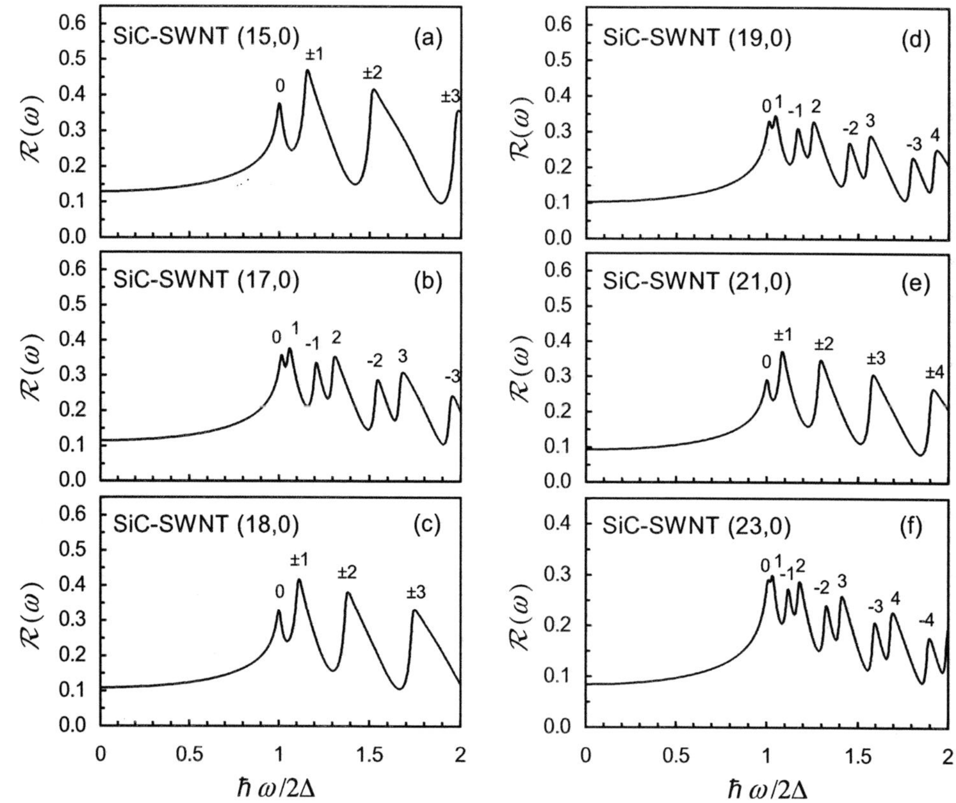

Figure 8. Calculated spectra of the reflectance $\mathcal{R}(\omega)$ for the same SiC-SWNT ensembles, as in Fig. 5. Resonant interband electron transitions are indicated by the numbers on the top of the peaks, as in Fig. 5.

Fig. 6. Indeed, as seen from this figure, the most prominent peak in the $\varepsilon_2(\omega)$ spectra always corresponds to the $v_1 \to c_1$ transition. However, in the $\varepsilon_2(\omega)$ spectra of the type-I-nanotube ensembles, there is a nearest-neighboring peak corresponding to the $v_0 \to c_0$ transition, situated on the left-hand side of the central peak (Figs. 6a, 6c and 6e), whereas for the type-II-nanotube ensembles a small cusp (Fig. 6b) or a narrow shoulder (Figs. 6d and 6f) is observed on the left-hand wing of the central peak. Besides this, we again observe deep dip structures in the $\varepsilon_2(\omega)$ spectra of the type-I-nanotube ensembles and, in contrast, rather shallow ones in the $\varepsilon_2(\omega)$ spectra of the type-II-nanotube ensembles.

Finally, there is one more point to be noted regarding the spectra in Figs. 5 and 6. The amplitudes of all the peaks in these spectra become smaller as the diameters of the nanotubes become larger. This implies that, if one desires to attain a larger optical response of SiC–SWNT ensembles, one should preferably use those fabricated from small-diameter tubes.

The qualitative features of the $\varepsilon_1(\omega)$ and the $\varepsilon_2(\omega)$ spectra, indicated above, are directly reflected in the $n(\omega)$ and the $\mathcal{R}(\omega)$ spectra, which are shown in Figs. 7 and 8, respectively. In particular, the spectral dependence of $n(\omega)$ closely resembles that of $\varepsilon_1(\omega)$ except for the noticeable diminishing of the intensity of the dips (cf. Figs. 5 and 7). Nevertheless,

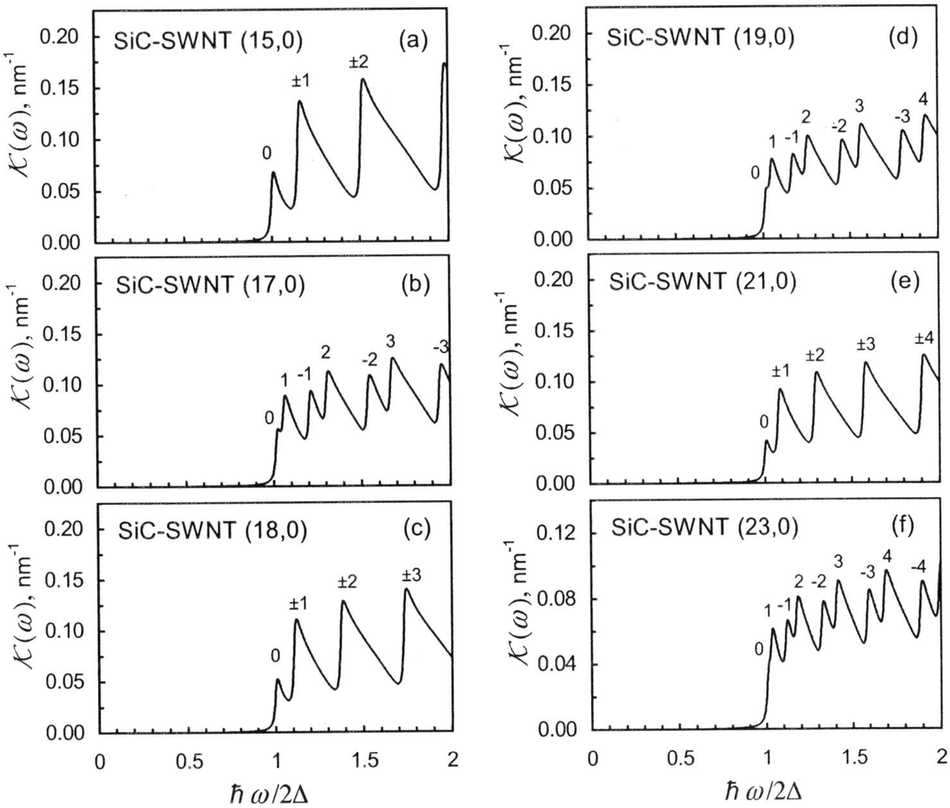

Figure 9. Calculated optical absorption spectra for the same SiC–SWNT ensembles, as in Fig. 5. Resonant interband electron transitions are indicated by the numbers on the top of the peaks, as in Fig. 5.

for the type-I-nanotube ensembles the refractive index $n(\omega)$ changes rather significantly in the vicinity of the resonances, the peak values of $n(\omega)$ being rather large near the threshold $\hbar\omega/2\Delta = 1$, especially for the smallest nanotubes considered (see Fig. 7a). It is well known (see, e.g., [50]), that the linear refractive index $n(\omega)$ enters the figure of merit of electro-optical materials. The above-mentioned spectral features of $n(\omega)$ for SiC–SWNTs imply that the well-ordered arrays of such tubes are possibly a promising material to be used in the future nanoscale optoelectronic devices, such as, for example, electro-optical modulators.

Turning next to Fig. 8, which displays the reflectance spectra $\mathcal{R}(\omega)$, we find the characteristic peak and dip structures, the spectral position of the former being determined by the energetic spacing between mirror pairs of van Hove singularities in the electronic density of states of both the above-mentioned types (I and II) of SiC–SWNTs. As seen from Fig. 8, the details of the behavior of $\mathcal{R}(\omega)$ depend on the types of the nanotubes, but the overall trend is that the reflectance $\mathcal{R}(\omega)$ decreases with increasing the nanotube radius. It is also worth noting that reflectance measurements may be an important tool to study the optical properties of SiC–SWNT samples in the spectral region below the threshold $\hbar\omega/2\Delta = 1$, which is inaccessible to the optical absorption spectroscopy.

Finally, we turn to a discussion of the optical absorption spectra shown in Fig. 9, which exhibit the characteristic resonances associated with the electron transitions between the occupied valence-band states and the unoccupied conduction-band ones (see Fig. 2). It is interesting to note that the spectral profile of the resonance lines in Fig. 9, as well as the relative strength of the resonances are markedly different from those in the $\varepsilon_2(\omega)$ spectra (Fig. 6), reflecting the influence of the refractive-index dispersion on the absolute strength and the shape of the optical resonances. In particular, the resonances observed in the $\mathcal{K}(\omega)$ spectra (Fig. 9) are apparently more broadened and asymmetrical that those in the $\varepsilon_2(\omega)$ spectra (Fig. 6), exhibiting a saw-tooth-like behavior. This is a direct consequence of rather broad dip structures in the $n(\omega)$ spectra (see Fig. 7). Furthermore, the presence of the quantity $n(\omega)$ in the denominator of Eq. (56) for the absorption coefficient $\mathcal{K}(\omega)$ reverses the relative intensity of the second and the subsequent peaks in the $\mathcal{K}(\omega)$ spectra as compared to their relative intensity in the $\varepsilon_2(\omega)$ spectra (cf. Fig. 9 and Fig. 6). Thus, our main prediction regarding the optical absorption spectra of the uniform ensembles of the zigzag SiC–SWNTs is that these spectra exhibit sharp characteristics features with an incremental increase of the peak intensity in varying the photon energy $\hbar\omega$ from the threshold value $\hbar\omega/2\Delta = 1$ to higher values. An experimental observation of such features, if realized, might be of primary importance for testing the band-structure model proposed here for the zigzag SiC–SWNTs.

5. Conclusion

To conclude, in this work we have put forward a theory for the linear optical properties of zigzag single-walled SiC nanotubes. To the best of our knowledge, no optical measurements on SiC nanotubes (single or bundles) have been made so far. We hope that the work presented in this chapter will stimulate such experiments, which could provide valuable information concerning the underlying electronic structure of these newly synthesized nanotubes and, in addition, open up a new promising field of scientific research and technological applications.

Acknowledgment

The work of Vl.A.M. and E.A.G. has been supported by the Russian Foundation for Basic Research.

References

[1] S. Iijima, "Helical microtubules of graphitic carbon", *Nature (London)* **354**, 56-58 (1991).

[2] M.S. Dresselhaus, G. Dresselhaus, and P.C. Eklund, *Science of Fullerenes and Carbon Nanotubes*, (Academic Press, San Diego, 1996).

[3] R. Saito, G. Dresselhaus, and M.S. Dresselhaus, *Physical Properties of Carbon Nanotubes*, (Imperial College Press, London, 1998).

[4] *Science and Application of Nanotubes*, edited by D. Tomanek and R.J. Enbody, (Kluwer Academic / Plenum, Dordrecht, 2000).

[5] *Carbon Nanotubes: Synthesis, Structure, Properties, and Applications*, edited by M.S. Dresselhaus, G. Dresselhaus, and Ph. Avouris, Springer Topics in Applied Physics, Vol. 80, (Springer, Berlin, 2001).

[6] S. Reich, C. Thomsen, and J. Maultzsch, *Carbon Nanotubes*, (Wiley-VCH, Weinheim, 2003).

[7] N.G. Chopra, R.J. Luyken, V.H. Crespi, M.L. Cohen, S.G. Louie, and A. Zettl, "Boron nitride nanotubes", *Science* **269**, 966-267 (1995).

[8] L. Boulanger, B. Andriotis, M. Cauchetier, and F. Willaime, "Concentric shelled and plate-like graphitic boron nitride nanoparticles produced by CO_2 laser pyrolysis", *Chem. Phys. Lett.* **234**, 227-232 (1995).

[9] A. Loiseau, F. Willaime, N. Demoncy, G. Hug, and H. Pascard, "Boron nitride nanotubes with reduced numbers of layers synthesized by arc discharge", *Phys. Rev. Lett.* **76**, 4737-4740 (1996).

[10] M. Terauchi, M. Tanaka, K. Suzuki, A. Ogino, and K. Kimura, "Production of zigzag-type BN nanotubes and BN cones by thermal annealing", *Chem. Phys. Lett.* **324**, 359-364 (2000).

[11] R.S. Lee, J. Gavillet, M. Lamy de la Chapelle, A. Loiseau, J.-L. Cochon, D. Pigache, J. Thibault, and F. Willaime, "Catalyst-free synthesis of boron nitride single-wall nanotubes with a preferred zig-zag configuration", *Phys. Rev. B* **64**, 121405 (2001).

[12] Q. Wu, Z. Hu, X. Wang, Y. Lu, X. Chen, H. Xu, and Y. Chen, "Synthesis and characterization of faceted hexagonal aluminum nitride nanotubes ", *J. Am. Chem. Soc.* **125**, 10176-10177 (2003).

[13] J. Goldberge, R. He, Y. Zhang, S. Lee, H. Yan, H. Chol, and P. Yang, "Single-crystal gallium nitride nanotubes", *Nature (London)* **422**, 599-602 (2003).

[14] R. Tenne, L. Margulis, M. Genut, and G. Hodes, "Polyhedral and cylindrical structures of tungsten disulphide", *Nature (London)* **360**, 444-446 (1992).

[15] L. Margulis, G. Solitra, and R. Tenne, "Nested fullerene-like structures", *Nature (London)* **365**, 113-114 (1993).

[16] Y. Feldman, E. Wasserman, D.J. Srolovitch, and R. Tenne, "High-rate, gas-phase growth of MoS_2 nested inorganic fullerenes and nanotubes ", *Science* **267**, 222-225 (1995).

[17] F. Krumeich, H.-J. Muhr, M. Niederberger, F. Bieri, B. Schnyder, and R. Nesper, "Morphology and topochemical reactions of novel vanadium oxide nanotubes", *J. Am. Chem. Soc.* **121**, 8324-8331 (1999).

[18] H. Imai, Y. Takei, K. Shimizu, M. Matsuda, H. Hirashima, and R. Nesper, "Direct preparation of anatase TiO_2 nanotubes in porous alumina membranes", *J. Mater. Chem.* **9**, 2971-2972 (1999).

[19] C.N.R. Rao and M. Nath, "Inorganic nanotubes", *Dalton Trans.*, 1-24 (2003).

[20] Z. Weng-Sieh, K. Cherrey, N.G. Chopra, X. Blase, Y. Miyamoto, A. Rubio, M.L. Cohen, S.G. Louie, A. Zettl, and R. Gronsky, "Synthesis of $B_xC_yN_z$ nanotubules ", *Phys. Rev. B* **51**, 11229-11232 (1995).

[21] C. Pham-Huu, N. Keller, G. Ehret, and M.J. Ledoux, "The first preparation of silicon carbide nanotubes by shape memory synthesis and their catalytic potential", *J. Catal.* **200**, 400-410 (2001).

[22] X.-H. Sun, C.-P. Li, W.-K. Wong, N.-B. Wong, C.-S. Lee, S.-T. Lee, and B.-K. Teo, "Templating effect of hydrogen-passivated silicon nanowires in the production of hydrocarbon nanotubes and nanoonions via sonochemical reactions with common organic solvents under ambient conditions", *J. Am. Chem. Soc.* **124**, 14464-14857 (2002).

[23] M.A. Lienhard and D.J. Larkin, *Silicon Carbide Nanotube Synthesized* (2002). http://www.grc.nasa.gov/WWW/RT2002/5000/5510lienhard.html.

[24] K. Nicolas, P.H. Cuong, E. Gabrielle, K. Valerie, and J.L. Marc, "Synthesis and characterisation of medium surface area silicon carbide nanotubes", *Carbon* **41**, 2131-2139 (2003).

[25] T. Taguchi, N. Igawa, H. Yamamoto, and S. Jitsukawa, "Synthesis of silicon carbide nanotubes", *J. Am. Ceram. Soc.* **88**, 459-461 (2005).

[26] A. Huczko, M. Bystrzejewski, H. Lange, A. Fabianowska, S. Cudzilo, A. Panas, and M. Szala, "Combustion synthesis as a novel method for production of 1-D SiC nanostructures", *J. Phys. Chem. B* **109**, 16244-16251 (2005).

[27] H. Wang, X.-D. Li, T.-S. Kim, and D.-P. Kim, "Inorganic polymer-derived tubular SiC arrays from sacrificial alumina templates", *Appl. Phys. Lett.* **86**, 173104 (2005).

[28] R.Q. Zhang, S.-T. Lee, C.-K. Law, W.-K. Teo, and B.-K. Teo, "Silicon nanotubes: Why not?", *Chem. Phys. Lett.* **364**, 251-258 (2002).

[29] A. Mavrandonakis, G. Froudakis, M. Schnel, and M. Mühlhäuser, "From pure carbon to silicon-carbon nanotubes: An ab-initio study", *Nano Lett.* **3**, 1481-1484 (2003).

[30] M. Menon, E. Richter, A. Mavrandonakis, G. Froudakis, and A.N Andriotis, "Structure and stability of SiC nanotubes", *Phys. Rev. B* **69**, 115322 (2004).

[31] M. Zhao, Y. Xia, F. Li, R.Q. Zhang, and S.-T. Lee, "Strain energy and electronic structures of silicon carbide nanotubes: Density functional calculations", *Phys. Rev. B* **71**, 085312 (2005).

[32] *Properties of Silicon Carbide*, edited by G.L. Harris, (INSPEC, Institution of Electrical Engineers, London, 1995).

[33] H. Ajiki and T. Ando, "Electronic states of carbon nanotubes", *J. Phys. Soc. Jpn.* **62**, 1255-1266 (1993).

[34] T. Ando, "Theory of electronic states and transport in carbon nanotubes", *J. Phys. Soc. Jpn.* **74**, 777-817 (2005).

[35] P. Král, E.J. Mele, and D. Tománek, "Photogalvanic effects in heteropolar nanotubes", *Phys. Rev. Lett.* **85**, 1512-1515 (2000).

[36] E.J. Mele and P. Král, "Electric polarization of heteropolar nanotubes as a geometric phase", *Phys. Rev. Lett.* **88**, 056803 (2002).

[37] R. Saito, M. Fujita, G. Dresselhaus, and M.S. Dresselhaus, "Electronic structure of graphene tubules based on C_{60}", *Phys. Rev. B* **46**, 1804-1811 (1992).

[38] R.A. Jishi, D. Inomata, K. Nakao, M.S. Dresselhaus, and G. Dresselhaus, "Electronic and lattice properties of carbon nanotubes", *J. Phys. Soc. Jpn.* **63**, 2252-2260 (1994).

[39] M.S. Dresselhaus, and P.C. Eklund, "Phonons in carbon nanotubes", *Adv. Phys.* **49**, 705-814 (2000).

[40] R. Rurali, "Electronic and structural properties of silicon carbide nanowires", *Phys. Rev. B* **71**, 205405 (2005).

[41] E.I. Blount, "Formalism of band theory" in *Solid State Physics, Advances in Research and Applications*, edited by F. Seitz and D. Turnbull, (Academic Press, New York, 1962), Vol.13, 305-373.

[42] E.M. Lifshitz and L.P. Pitaevskii, *Statistical Physics*, Part 2, (Pergamon Press, New York, 1980), Chap. 6.

[43] H. Ajiki and T. Ando, "Aharonov-Bohm effect in carbon nanotubes", *Physica B (Amsterdam)* **201**, 349-352 (1994).

[44] H. Ajiki and T. Ando, "Carbon nanotubes: optical absorption in Aharonov-Bohm flux", *Jpn. J. Appl. Phys.* **34**, Suppl. 34-1, 107-109 (1995).

[45] V.M. Genkin and P.M. Mednis, "On the theory of nonlinear effects in crystals in which partially filled bands are taken into account", *Zh. Eksp. Teor. Fiz.* **54**, 1137-1150 (1968) [*Sov. Phys. JETP* **27**, 609-618 (1968)].

[46] Vl.A. Margulis and E.A. Gaiduk, "Nature of near-infrared absorption in single-wall carbon nanotubes", *Phys. Lett. A* **281**, 52-58 (2001).

[47] Vl.A. Margulis and E.A. Gaiduk, "Dielectric function of single-wall carbon nanotubes", *Chem. Phys. Lett.* **341**, 16-22 (2001).

[48] A. Gali, "*Ab initio* study of nitrogen and boron substitutional impurities in single-walled SiC nanotubes", *Phys. Rev. B* **73**, 245415 (2006).

[49] S. Tatsuura, M. Furuki, Y. Sato, I. Iwasa, M. Tian, and H. Mitsu, "Semiconductor carbon nanotubes as ultrafast switching materials for optical telecommunications", *Adv. Mater. (Weinheim, Ger.)* **15**, 534-537 (2003).

[50] T.S. Moss, G.J. Burrell, and B. Ellis, *Semiconductor Opto-Electronics*, (Butterworth, London, 1973).

In: Pattern Recognition in Nanoscience...
Editor: Diane M. Perez, pp. 123-157

ISBN: 1-60021-715-X

Chapter 5

PATTERN RECOGNITION METHODS IN ENVIRONMENTAL RADIOACTIVITY STUDIES

***S. Dragović*[1] *and A. Onjia*[2]**

[1]Institute for the Application of Nuclear Energy - INEP, Banatska 31b, 11080 Belgrade, Serbia

[2]The Vinča Institute of Nuclear Sciences, P.O. Box 522, 11001 Belgrade, Serbia

Abstract

Pattern recognition methods provide powerful tools for the analysis and interpretation of large environmental data sets generated within environmental monitoring programmes. Most of these data sets consist of trace elements and/or trace organic pollutants patterns. Only a few studies have been done on employing the pattern recognition methods to describe the correlation between locations and radioactive contamination. This multivariate approach has been used predominantly for the identification of radioactive isotopes, quantitative gamma-ray spectrometry analysis and for optimization of gamma-ray spectrometric measurements. Spatial data analysis based on radioactive contamination of diverse regions is complex for many reasons. These include the uncertainty of the source term, high spatial and temporal variability of pollution patterns, spatial and temporal non-stationary and the multivariate nature of the phenomenon with linearly and non-linearly correlated variables. Our studies show that the geographic origin can be recognized with minimum effort if the relevant constituents are analyzed and the results are included in data analysis algorithms. Five common pattern-recognition techniques, artificial neural networks (ANN), principal component analysis (PCA), linear discriminant analysis (LDA), k-nearest neighbors (kNN) and soft independent modeling of class analogy (SIMCA) were employed to classify soil and bioindicator samples (mosses and lichens) according to their geographical origin, based on their content of radionuclides from different sources (members of the natural uranium and thorium decay chains, cesium isotopes originating from the Chernobyl power plant accident and cosmogenic beryllium), determined by gamma-ray spectrometry. The ability of the ANN to extract hidden features from the input signals was found to be particularly useful for this kind of monitoring, when the data sets with complex correlation structures had to be analyzed, or when data sets contained series of many highly inter correlated variables.

1. Introduction

Environmental monitoring studies generate large amounts of data particularly concerning contaminants distributed throughout distinct geographical sites during different time periods. The detection of spatial structures, the assessment of polluted areas and their loading state, as well as the evaluation of the sampling plan are essential parts of environmental questions. Methods of univariate statistics can successfully be applied if a clear spatial structure of the pollutant exists and the sampling plan was well organized to detect this structure. The most commonly occurring problems when dealing with contamination data are multivariate and multiscale structures of data, high spatial variability, presence of noise and outliers. Spatial trends limit application of stationary geostatistical models. Multivariate chemometric techniques offer an alternative way of dealing with spatial information when no stationarity of the model is assumed. If the data set contains different concentrations of different pollutants measured at different sampling locations, which is a very frequent case in soil pollution studies, methods of multivariate statistics must be applied (Einax and Soldt, 1999). These techniques reveal similarities in the data set and can help to locate existing pollution patterns. It is possible to detect main emission sources and to uncover the mechanisms in the environmental compartment under investigation. The advantage is that the whole information of the variance covariance matrix can be treated simultaneously.

Chemometric data analysis methods provide powerful tools for the analysis and interpretation of large, environmental, multivariate data sets generated within environmental monitoring programs (Massart et al., 1998; Einax et al., 1997). The basic assumptions of these multivariate exploratory data studies is that main sources of data variance, observed in the concentration changes of contaminants, are due to reduced number of contamination sources of different origin which are defined by profiles describing their chemical composition and temporal distributions. Pattern recognition methods have been used in chemometrics to reveal and evaluate complex relationships in a wide variety of environmental applications. Common uses of these techniques are to identify factors that are combinations of measurable variables, to illustrate groups or cluster associations among samples, to assess spatial distribution of environmental factors or perturbations and to predict a property of interest (such as biological response to chemical perturbation).

Pollution modelling is generally pointed toward the identification of man-made sources of contaminants. By understanding the spatial and temporal variations of these pollutants, control measures can be applied to bring levels into compliance with environmental standards. Multivariate chemometric modelling techniques can help in situations where the emission of pollutants is added to natural sources of same materials. Multivariate classification and calibration methods are particularly well-suited to extract predictive information from a set of measurements that individually may show only a small correlation to the property of interest. Through the use of multivariate models we can be more effective in charting relationships between pollution variables. Chemometric assessments have also been used to correlate environmental factors to specific human health concerns. The application of chemometrics in environmental analysis can result in detection of pollution contributions from a complex mixture of sources, assessment of geographical or atmospheric distributions and influences, prediction of biological response to perturbation, understanding the interplay of influential factors which can not be directly measured, optimization of processes for controlling plant waste or recycling and improvement in the interpretability of analytical instrument data.

Ecological modelling studies have traditionally been concentrated on the use of differential equation models (Jorgensen, 1976; Ross, 1976). While this models provide valuable insight into the particular ecosystem studied, their strong dependence on the parameters specific to that system made these models prone to criticism as being case-specific. To achieve generalizability and thus to gain insight into the ecosystem in general, empirical models have been one of the main practices (Tan and Beklioglu, 2005; Özesmi et al., 2005). This is especially true for statistical classification and pattern recognition models which fundamentally aim to characterize a combination of variables and their measurements which lead to particular state(s) of the ecosystem.

Traditional classification models require strong assumptions about the distribution of the underlying observations and/or the system characteristics (Hastie et al., 2001). Nonlinear and/or nonparametric models overcome this issue with finite-size observations. These models are able to capture the nonlinear interactions and complexity of the ecosystem without the need for any assumptions about the distribution of the observations.

In recent years the analysis and processing of spatially distributed and time dependent data has become a very important subject due, on one hand, to comprehensive development of environmental and pollution monitoring networks even leading to data mining problems and, on the other hand, to much better understanding of data driven machine learning models.

The pattern recognition problem can be divided into two main categories: (i) supervised classification: where the state of nature for each pattern is known and (ii) unsupervised classification: where the state of nature is unknown and learning is based on the similarity of patterns (Jain et al., 2000). A supervised classification could be subdivided into two main phases, namely the training phase and testing phase. In the training phase, the classifier is made to learn by known categories of patterns and in the classification or the testing phase, unknown patterns, which were not part of the training dataset, are assigned class label of the nearest category of trained patterns.

How well a given pattern classifier/recognizer can classify or make some predefined decisions in the shortest possible time and at lowest cost would determine its performance. In general, the performance of a classifier depends on several factors. Some of them are (Jain et al., 2000):

i) number of training samples available to the classifier,
ii) generalization ability, i.e. its performance in classifying test patterns which were not used during the training stage,
iii) classification error - some measured value based on the incorrect decision of the class labelling of any given pattern,
iv) complexity - in some cases the number of features or attributes (dimensions) are relatively larger than the number of training samples, usually referred as curse of dimensionality,
v) speed - processing speed of training and/or testing phase(s),
vi) storage - total amount of parameters required to store after the training phase for classification (testing) purposes (Sharma et al., 2006).

Multivariate analysis methods have succesfully been used in biomonitoring of environmental contamination. The amount of information extracted from data obtained in biomonitoring programmes by univariate methods is only a fraction of that contained within. The full potential of biological monitoring can only be achieved through the development of more advanced data interpretation methods, such are multivariate methods. The main advantages of these methods over traditional mthods are that: they can be used for classification or for detailed assessment of environmental quality, they do not require the establishment of a reference state, they are not founded on any existing subjectively derived biomonitoring system and they are holistic in terms of its pattern recognition process (Walley and O'Connor, 2001). One of the major advantages of the pattern recognition approach to the interpretation of bioindicator data is that, unlike the existing biomonitoring systems, it is not founded on subjective human evaluations. In biomonitoring studies, different multivariate statistical analysis techniques have been used in order to obtain information on the character and the source of pollution from elemental concentration patterns obtained from moss and lichen data sets (Hopke, 1988; Sloof and Wolterbeek, 1991). The bioindication ability of different species of mosses and lichens with respect to the accumulation of contaminants in various locations has also been investigated using these techniques (Stratis et al., 1996; Tsakovski et al., 1999; Stratis et al., 1998). The spatial pattern of metal content in mosses and lichens has been detected and discussed in relation to local contamination sources (Figuera et al., 2002).

The range of multivariate techniques are used for predictive modelling in environmental monitoring studies, from regression to discriminant analysis and from generalized additive and linear models to neural networks (Tan et al., 2006).

Spatial data analysis of radioactively contaminated areas are particularly challenging for many reasons: uncertainty of the source term, high spatial and temporal variability of pollution patterns, spatial and temporal nonstationarity and multivariate nature of the phenomenon with linearly and nonlinearly correlated variables. The chemometric approach was used to describe the correlation between locations and radioactive contamination (Kanevski et al., 1996; Kanevski et al., 1997; Dragović and Onjia, 2007a). The problem of radioactive pollution transport is essential for analysing nuclear accident consequences, possible risks and countermeasures in a case of release of radioactive materials. The interest to radioactive transport modelling and exposure dose estimation was raised after the accident at the Chernobyl Nuclear Power Plant, which lead to ecological, social, environmental and economical consequences. The number of models to assess the problem of large scale nuclear accident, monitoring of contaminated territories and population have been developed worldwide. Methods of geostatistics and stochastic modelling were used to analyse spatial distribution of radionuclides in the Chernobyl fallout and spatial radioecological pollution information in general. Statistical approach was originally proposed forty years ago to solve prediction problems in mining. The main assumption of statistical approach is that the explored phenomenon is considered as a random function, and the data samples are its realisations. This does not mean that the phenomenon is random, but it allows us to take into account variability of the spatial process, which is reflected by the structure of measurement distribution. The key role in statistical approach plays spatial correlation between the samples. It is explored and modelled using variograms - second-order moment of distributions. Spatial structure model can be further used in different regression algorithms or in stochastic simulation algorithms, which model spatial variability of the phenomenon. Statistical methods show good performance in different applications. However, they require stationary assumption in a certain form that is sometimes difficult to reach especially in the applications dealing with environmental data. Prediction and risk mapping has been carried out using both traditional deterministic interpolators and geostatistical regression models. Stochastic modelling techniques have been applied to model high spatial variability of the contamination pattern. Hybrid geostatistical-neural network models have been developed to improve spatial predictions in case of strong spatial trends and non-stationary.

Artificial neural networks (ANNs) are one of the most frequently used multivariate statistical methods in environmental radioactivity studies. This approach has been used for identifying of radioactive isotopes (Olmos et al., 1991; Olmos et al., 1994), quantitative gamma-ray spectrometry analysis (Vigneron et al., 1996; Pilato et al., 1999) and for optimization of gamma-ray spectrometric measurements (Dragović et al, 2005; Dragović and Onjia, 2005).

The ANNs were suggested as viable solutions to diagnostic and fault identification issues in nuclear power plants a few decades ago (Yiftah, 1988). They were used to classify abnormal reactor internal state using neutron noise spectra analysis (Frean and Robins, 1977) and for identification of a nuclear power plant component fault by analysis of the resulting transient data (Reifman and Vitela, 1994).

Compact portable systems capable of quickly identifying contaminants in the field are of great importance when monitoring the environment. One of the missions of the Pacific Northwest Laboratory is to examine and develop new technologies for environmental restoration and waste management at the Hanford Site (a former plutonium production facility). This effort includes the development of portable, inexpensive systems capable of real-time identification of contaminants in the field. Radiation sensor system, which contains a portable gamma-ray spectrometer, was developed and used to identify and quantify radioactive isotopes (Keller et al., 1994). The ANN approach was employed for pattern recognition on the entire spectrum, i.e. the whole spectrum was used in the identification process instead of individual peaks in the spectrum. Operation consisted of presenting an unknown sample to the system, generating a gamma-ray spectrum, and passing the spectrum to the ANN which produces a classification of the unknown sample. The values of the output neurons were proportional to the quantities of each radioactive isotope found in the sample. These prototypes also demonstrated several advantages of this approach over conventional analytical techniques including compactness, portability, real-time analysis, and automation. Each isotope presented to the spectrometer produced a spectrum that is characteristic of that isotope. By presenting many different isotopes to the system, a database of spectra was constructed. From this database, training sets and test sets were generated. These sets were collections of labeled patterns (spectra of isotopes with known concentrations), representative of the desired identification mapping. The training sets were used to configure the ANN. The goal of this training was to learn an association between the spectra and the labels presenting the spectra.

The ANNs were used for improvement of the energy spectrum characteristics of a CdZnTe semiconductor detector, based on a fusion method that uses both digital and analog information (Mardiyanto et al., 2000). A digital and analog system were simultaneously applied for the detector output. The digital system digitized the signal pulse shapes for pattern recognition based on a neural network algorithm. The analog measured the signal pulse heights with a conventional analog system. The digital and the analog output data were fused and then used for improvement of the energy spectrum characteristics, i.e. the photopeak area, the peak-to-background ratio, and the FWHM of photopeaks.

ANN technique was also applied in the field of nuclear spectroscopy, e.g. for resolution improvement of CdZnTe detectors employing pulse-shape treatment (Sakai et al, 1999) and X-, gamma- and alpha-ray spectrometry (Kangas et al., 1994; Keller and Kouzes, 1994; Keller et al., 1995). The ANN method was successfully applied to the analysis of gamma-ray spectra composed of mixed radioisotopes and spectra of uranium ores (Yoshida et al., 2002). It has been shown that analysis of complex spectra of fission and activation products can be correctly handled by ANN and the procedure can be automated with minimum laboratory measurements for network training (Pilato et al., 1999). The method allowing to associate an uncertainty to each neural network output was also developed. Such a method can be applied to online systematic analysis, where a precise result is required in a very short time, without experienced user intervention.

ANN approach has been applied to the analysis of photon spectra from uranium-enrichment measurements (Vigneron et al., 1996). It has been shown that this approach appears to be useful when a fast response is required with reasonable accuracy, no spectral hypothesis are made, and no definitive mathematical model can be assigned a priori. Final network with connections and weighting functions could be easily implemented using commercial digital processing hardware. There is no single learning procedure which is appropriate for all tasks. It is of fundamental importance that special requirements of each task are analyzed and that appropriate training algorithms are developed for families of tasks.

In this paper, we will give an overview of recent uses of pattern recognition methods in environmental radioactivity studies, either for optimization of measurement conditions or for classification of environmental samples according geographical origin based on their radionuclide content.

2. Multivariate Analysis Methods

2.1. Artificial Neural Network with Back-Propagation Learning

The training process in back-propagation networks is done in two phases: feed-forward and back-propagation (Werbos, 1974; Rumelhart et al., 1986). In the feed-forward phase, the input layer neurons pass the input values on to the hidden layer. Each of the hidden layer neurons computes the weighted sum of its inputs passes the sum through its activation function and presents the activation value to the output layer. After computation of the weighted sum of each neuron in the output layer, the sum is passed through its activation function, resulting in one output value for the network. The frequently used activation function in neural network architectures is a sigmoidal function defined as:

$$o_j = f(net_j) = 1/\{1+\exp[-(net_j+\theta_j)]\} \quad (1)$$

where θ is a bias term or threshold value of the j node responsible for accommodating non-zero offsets in the data.

In the back-propagation phase, the error between the network output and the desired output values is calculated using the so-called generalized delta rule and weights between neurons are updated from the output layer to the input layer as follows:

$$w_{ji}^{n+1} = w_{ji}^{n} + \eta\delta_j o_j + \alpha w_{ji}^{n} \qquad (2)$$

where δ_j is the error signal at neuron j, o_j is the output of neuron j, n is the number of iterations, and η and α are learning rate and momentum, respectively. The learning rate controls the rate at which the network learns. The momentum term has the effect of adding a proportion of the previous weight change during training (Maier and Dandy, 1998). The training process is successfully completed when the iterative process has converged.

In order to optimize model performance, it is vital to know at what count a local minimum in the error surface is reached and what the magnitude of the oscillations in the forecasting error will be if training is continued. This requires three data sets: a training set, used to train the network, a test set, used to evaluate the generalization ability of the network and a validation set, used to assess the performance of the model once the training phase has been completed (the process called cross-validation).

The "leave-10%-out" is a frequently used method for cross-validation. With this method, 10% of the data in the training set are not used for updating of weights, so this 10% can be used as an indication of whether or not memorization is taking place. When an ANN memorizes the training data, it produces acceptable results for the training data, but poor results when tested on unseen data.

One pass through the set of training patterns along with the updating of weights is called the cycle or epoch. In the testing phase, the trained neural network is fed by a separate set of data. In this phase, the neural network predictions (using the trained weights) are compared with the target output values. The neural network will produce almost instantaneous results of the outputs for practical inputs provided. Once the training and testing phases are found successful, the neural network can be put to use in practical applications. The root mean square error (RMSE) was used to estimate the success of the training process. It is computed as

$$RMSE = [\sum_{i=1}^{n} (d_i - o_i)/n]^{1/2}/x \qquad (3)$$

where d_i is the desired output (experimental values) in the training or testing set, o_i is the actual output (ANN predicted values) in the training or testing set, n is the amount of data in the training or testing set and x is the average value of the desired output in the training or testing set.

The different training algorithms are used in neural networks with back-propagation learning: online back-propagation (OBP), online back-propagation randomized (OBPR), batch back-propagation (BBP), resilient propagation (RP) and quick propagation (QP) (QwikNet Version 2.23, 1999). The idea of the back-propagation algorithms is to propagate the error back through the network and to adjust the weights of each layer as it propagates. OBP updates the weights after each pattern is presented to the network. In the OBPR algorithm the order of input patterns is randomized prior to each epoch which makes the learning process stochastic. In the BBP algorithm, weight updates occur after each epoch. RP is an adaptive learning rate method where weight updates are based only on the sign of the local gradients, not on their magnitude. Each weight has its own updated value which varies with time. QP is a training method with a fixed learning rate that needs to be chosen to suit the problem.

2.2. Principal Component Analysis

Principal component analysis is a powerful, linear pattern recognition technique used as a mathematical tool for analyzing, classifying and reducing the dimensionality of numerical datasets in a multivariate problem (Brereton, 2003; Joliffe, 1986; Antonić et al., 2003; Hopke, 2003). Generally, the k-th principal component PC_k, is a linear combination of the n response vectors X_{ij} for the analysis under study:

$$PC_k = a_{1k}X_{1j} + a_{2k}X_{2j} + a_{3k}X_{3j} + ... + a_{nk}X_{nj} \quad (4)$$

whilst the sum of the coefficients a_{ik}, called loadings, is set to unity. The loading of a single variable indicates how much that variable participates in defining the PC (the squares of the loadings indicate their percentage in the PC).

Typically, PCA decomposes the primary data matrix by projecting the multidimensional dataset onto a new coordinate base formed by the orthogonal directions with maximum variance in the data. The eigenvectors of the data matrix are called principal components and they are mutually uncorrelated. The principal components PC_k are ordered so that PC_1 displays the greatest amount of variance, followed by the next greatest PC_2 and so on. The magnitude of each eigenvector is expressed by its own eigenvalue, which gives a measure of the variance related to that principal component. As a result of the coordinate change, it is possible to achieve a data dimensionality reduction to the most significant principal components and elimination of the less important ones without any considerable information loss. In mathematical terms, n correlated random variables are transformed into a set of $d \leq n$ uncorrelated variables. These uncorrelated variables are linear combinations of the original variables and can be used to express the data in a reduced form. The main features of PCA are the coordinates of the data in the new base (score plot) and the contribution to each component of the variable (loading plot). The score plot is usually used for studying the classification of the data clusters; while the loading plots can be used for giving information on the relative importance of the variable to each principal component and their mutual correlation (Penza et al., 2002). In other terms, the PCA procedure can be applied by finding the eigenvectors of the primary matrix of the data points, and to form a transformation matrix from these eigenvectors ordered, so that the corresponding eigenvalues are in decreasing order. The number of significant components is estimated from the values of eigenvalues.

The technique has three effects: it orthogonalizes the components of the input vectors (so that they are uncorrelated with each other), it orders the resulting orthogonal components (principal components) so that those with largest variation come first, and it eliminates those components which contribute least to the variation in the data set (Wu and Massart, 1996).

2.3. Linear Discriminant Analysis

Linear discriminant analysis (LDA) is a linear and parametric method for feature reduction with discriminating character (Coomans et al., 1978; Massart et al., 1998a). The basic theory of LDA is to classify objects by dividing a high-dimensional descriptor space into two regions that are separated by a hyper plane defined by a linear discriminant function. LDA aims to find optimal discriminant features by maximizing the ratio of the between-class distance to the within-class distance of a given data set under supervised learning conditions. It finds the optimal transformation matrix in such a way to preserve most of the information that can be used to discriminate between the different classes. The features in the reduced space are linear combinations of the features in the original high dimensional space, where the coefficients of the linear combinations depend on the transformation matrix.

Assume there are k classes in the dataset. Suppose m_i, S_i, P_i are the mean vector, covariance matrix, and a priori probability of the i-th class, respectively, and m is the total mean. Then the between-class scatter matrix S_b, the within-class scatter matrix S_w, and the total scatter matrix S_t are defined as follows (Fukunaga, 1990):

$$S_w = \sum_{i=1}^{k} P_i \mathrm{S}_\mathrm{i} \tag{5}$$

$$S_b = \sum_{i=1}^{k} P_i (m_i - m)(m_i - m)^\mathrm{T} \tag{6}$$

$$S_t = S_b + S_w \tag{7}$$

For the covariance matrix S_i of the i-th class, we can decompose it as $S_i = X_i X_i^T$, where each column in X_i corresponds to a data point from the i-th class subtracted by its mean m_i.

If matrices are presented as

$$H_w = [\sqrt{P_1} X_1, \ldots, \sqrt{P_k} X_k] \tag{8}$$

$$H_b = [\sqrt{P_1}(m_1 - m), \ldots, \sqrt{P_k}(m_k - m)] \tag{9}$$

Then the scatter matrices S_w and S_b can be expressed as $S_w = H_w H_w^T$ and $S_b = H_b H_b^T$ and traces can be computed as follows:

$$\text{trace } (S_w) = \sum_{i=1}^{k} P_i \|X_i\|_\mathrm{F}^2 \tag{10}$$

$$\text{trace } (S_b) = \sum_{i=1}^{k} P_i \|m_i - m\|^2 \tag{11}$$

where $\|\cdot\|_F$ denotes the Frobenius norm (Golub and Van Loan, 1996). Hence, trace (S_w) measures the closeness of the vectors within classes, while trace (S_b) measures the separation between classes.

In the lower-dimensional space resulting from the linear transformation G, the within-class and between class matrices become

$$S_w^L = (G^T H_w)(G^T H_w)^T = G^T S_w G \tag{12}$$

$$S_b^L = (G^T H_b)(G^T H_b)^T = G^T S_b G \tag{13}$$

An optimal transformation G would maximize trace (S_b^L) and minimize trace (S_w^L). Classical LDA aims to compute the optimal G, such that

$$G = \arg\max_{G} \text{trace}\left((G^T S_w G)^{-1} G^T S_b G\right). \tag{14}$$

The solution can be obtained by solving an eigenvalue problem on $S_w^{-1} S_b$ (Fukunaga, 1990), provided that the within-class scatter matrix S_w is non-singular. Since the rank of the between-class scatter matrix is bounded from above by k-1, there are at most k-1 discriminant vectors by classical LDA. The requirement that one of the matrices must be non-singular restricts the application of LDA to datasets in which the dimension of the data does not exceed the sample size.

2.4. *k* -nearest Neighbour Algorithm

The k-nearest neighbour (kNN) algorithm is an instance based method, which assumes that all instances correspond to points in an n-dimensional space (Mitchell, 1997). Performance of a nearest neighbour classifier depends on the distance function and the value of the neighbourhood parameter k. Euclidean metric is the most popular choice for the distance function. If the measurement variables are not of comparable units and scales, it is more meaningful to standardize the variables before using the Euclidean distance for classification. Let an arbitrary instance X be described by the vector $[p_1(X), p_2(X), \ldots, p_{N_{par}}(X)]$, where $p_n(X)$ is the value of the n-th (parameter) of instance X, and N_{par} is the number of parameters of instance X. Then, the distance between two instances X_i and X_j is defined as d (X_i, X_j):

$$d(X_i, X_j) = \sqrt{\sum_{n=1}^{N_{par}} \left(p_n(X_i) - p_n(X_j)\right)^2} \tag{15}$$

The objective function of a new instance is evaluated according to the objective functions of its k-nearest neighbors. These functions may be in general either a discrete-valued or a real-valued function.

The kNN algorithm for approximating a discrete-valued objective function *f*, returns a value $\hat{f}(X_q)$ as an estimate of $f(X_q)$, by choosing the most common value of *f* among the *k* training samples nearest to X_q. If k=1, for example, then the 1-nearest neighbor algorithm assigns to $\hat{f}(X_q)$ the value of the objective function of the training sample nearest to X_q. For larger values of *k*, the algorithm assigns the most common value among the *k*- nearest training samples.

For continuous-valued objective functions, the algorithm calculates the mean value of the *k*-nearest training samples rather than calculating their most common value.

Thus, the objective function of a new instance is evaluated as:

$$\hat{f}(X_q) = \frac{\sum_{i=1}^{k} f(X_i)}{k} \quad (16)$$

The optimum value of k depends on the specific data set and it is to be estimated using the available training sample observations. In this context, the different validation methods are used. The most frequently used method is cross-validation. It is used to estimate the misclassification rate for different values of k and chooses that one which leads to the lowest estimate of misclassification rate.

2.5. Soft Independent Modelling of Class Analogy

The soft independent modelling of class analogy (SIMCA) is supervised classification technique which aims to create classification rules for each group of objects separately by constructing a confidence region around every group, using a set of model objects (Wold, 1976). To achieve this goal, after data pre-processing, the PCA model with *f* factors is constructed. It is done for each group of objects, by decomposing a sub-matrix X_k containing *m* objects of the *k*-th group, characterized by *n* explanatory variables. The PCA model for the k-th group, obtained from the singular value decomposition of X_k, can be presented as

$$\mathbf{X}_k - \overline{\mathbf{X}}_k = S_k \cdot V_k \cdot \mathbf{P}_k^T + E_k = T_k \cdot \mathbf{P}_k^T + E_k \quad (17)$$

where $\overline{\mathbf{X}}_k$ is the matrix containing in columns the mean of columns of $\mathbf{X}_k$, S_k is the score matrix containing *f* normalized score vectors in columns, V_k is the diagonal matrix with singular values, P_k matrix holds *f* loadings, T_k is the matrix of principal components, and E_k is the matrix of residuals.

To find an optimal number of factors (*f*) in the PCA model, a cross-validation procedure is usually carried out. The residuals of objects from the *k*-th group are assumed to follow a normal distribution. The residual standard deviation, σ_k, of the objects residuals from the *k*-th group can be computed either using the residuals of objects from the PCA model with *f* components or by considering the non-retained (*r*-*f*)-components, where *r* is the data rank:

$$\sigma_k = \sqrt{\sum_{i=1}^{m} \sum_{j=1}^{n} \frac{e_{ij}^2}{(r-f)(m-f-1)}} = \sqrt{\sum_{i=1}^{m} \sum_{j=f+1}^{r} \frac{t_{ij}^2}{(r-f)(m-f-1)}} \tag{18}$$

where e_{ij}^2 is the element of residual matrix E_k corresponding to the i-th sample and the j-th variable.

In order to construct a boundary around a group of objects, a confidence limit for the residuals of all objects, σ_{crit}, is determined with the F-test with nominator degrees of freedom $(r\text{-}f)$ and denominator degrees of freedom $(r\text{-}f)(m\text{-}f\text{-}1)$ and assumed significance level (De Maesschalck et al., 1999):

$$\sigma_{crit} = \sqrt{F_{crit}}\ \sigma_k \tag{19}$$

To check if a new object x_{new} belongs to the k-th group, the object is projected onto a principal component space constructed for objects of the k-th group and spanned by f principal components:

$$t_{new} = (\mathbf{x}_{new} - \overline{\mathbf{x}}_k)\, P_k \tag{20}$$

Then, the residual standard deviation, σ_{new}, is calculated for the new object:

$$\sigma_{new} = \sqrt{\sum_{i=f+1}^{r} \frac{t_{new,i}^2}{r-f}} \tag{21}$$

The residuals of new objects are compared to the residuals of all objects from the k-th group. A new object belongs to the k-th group if its residuals are smaller than residuals of the k-th group objects. As long as the values of σ_{new} are smaller than the critical value σ_{crit} for a given group, the new objects belong to this group. This condition implies that one object can be a member of more than one group, which explains why SIMCA is a soft classification technique.

3. Recent Applications of Pattern Recognition Methods in Environmental Monitoring Studies

Different applications of pattern recognition techniques in environmental monitoring of radioactive contamination, i.e. for optimization of gamma-ray spectrometric measurements and for classification of environmental samples according to geographic origin will be presented here. For all applications, the measurements were performed using a low-background HPGe gamma-ray spectrometer ORTEC-AMETEK (34% relative efficiency and 1.65 keV FWHM for ^{60}Co at 1.33 MeV, 8192 channels). Gamma Vision 32 software was used to process the spectra obtained (ORTEC, 2001). The following radionuclides detected in soil samples were analyzed: ^{226}Ra, ^{238}U, ^{235}U, ^{40}K, ^{232}Th, ^{134}Cs, ^{137}Cs and ^{7}Be.

3.1. Optimization of Gamma-Ray Spectrometric Measurements

One of the most frequently used pattern recognition techniques, artificial neural networks (ANNs), were employed to optimize important parameters in gamma-ray spectrometric analysis used for environmental monitoring: measuring uncertainty, peak-to-background ratio and minimum detectable activity. The ANN simulation was made using QwikNet software (QwikNet Version 2.23, 1999).

3.1.1. Optimization of Measuring Uncertainty

ANN was employed and optimized to model the measuring uncertainties of activities of radionuclides in natural soil samples (Dragović et al., 2005)[1]. The main aim of the study was to investigate the ability of a network, trained with a small set of experimental data, to provide a reliable estimate of the minimum measurement time needed to attain the desired average statistical accuracy in environmental samples of complex composition. This could be of special importance in the environmental monitoring programmes, when a large number of environmental samples have to be measured and when the time of measurement is important for the economization of experiments. It was shown that ANNs approach can determine a minimum measuring time which is requested for fulfilling quality assurance criteria for any particular application. It was also demonstrated that ANN, which is trained on random selected samples of the soil, is capable of predicting uncertainties for samples with different $^{238}U/^{232}Th$ ratio.

Since there are no theoretical principles for choosing the proper network topology, several different ANN structures were trained. ANNs were trained using different numbers of hidden nodes and learning epochs. When the network was optimized, the testing data were fed into the network to evaluate the trained network.

To determine the optimum number of hidden layer nodes, ANNs with different numbers of hidden layer nodes were trained. Finding the optimal number of hidden nodes is important since their function is to detect relationships between network inputs and outputs. The minimum error was obtained when the number of hidden layer nodes was equal to 6, so this value was chosen for optimization of the number of learning epochs. To select the most favorable number of learning epochs, RMSE values versus number of learning epochs were evaluated for both the training and testing sets. In both cases RMSE decreased rapidly at the beginning, and at 20000 epochs, the error of the testing set reached a minimum afterwards the test error began to increase. This effect is called overtraining or over fitting and inspection of the cross-validation set is crucial for detection of this effect. The critical moment in training is when the error of this set starts to increase while the training error is still decreasing. From that point, improvement of training error is achieved by better fitting of the experimental points rather than by fitting the general trend of the data. In order to avoid overtraining,

[1] Figures and tables in this section are reprinted from Dragović, S.; Onjia, A.; Stanković, S.; Aničin, I.; Bačić, G. *Nucl. Instr. Meth. Phys. Res. A* 2005, 540, 455-463, with kind permission from Elsevier.

20,000 learning epochs were selected as the optimal value. The optimized ANN is schematically represented in Fig. 1.

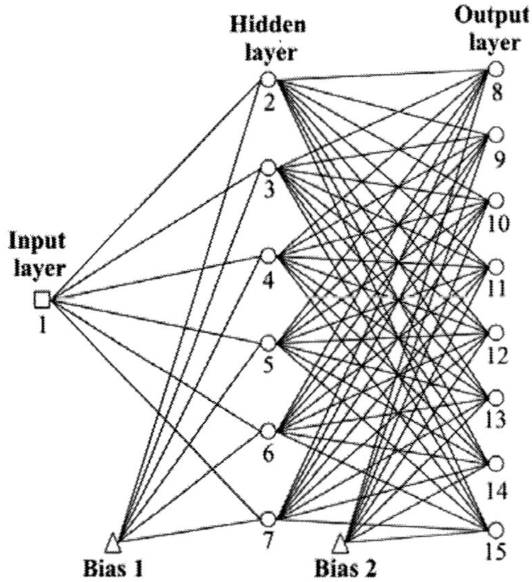

Figure 1. Schematic representation of a three-layer feed-forward neural network. Input layer node: 1, Hidden layer nodes: 2-7, Output layer nodes: 8-15.

The input layer consists of one node representing the time of measurement (t). The output layer consists of eight nodes representing the measuring uncertainties of the investigated radionuclides (^{226}Ra, ^{238}U, ^{235}U, ^{40}K, ^{232}Th, ^{134}Cs, ^{137}Cs and ^{7}Be). In addition, there are also a biases connected to nodes in the hidden (bias 1) and output (bias 2) layers via modifiable weighted connections. The biases are responsible for accommodating nonzero offsets in the data. They act like other processing elements that have a constant output. The purpose of this is to scale the input to a useful range.

The optimized neural network model was used to predict uncertainties of radionuclide activities for the remaining fourteen measurement times, i.e., those which were not used for the ANN training. The predictive power was assessed by comparing experimental and predicted uncertainties. The results show that the proposed neural network model adequately generalizes data and that it can be used for modeling of uncertainties. Correlation coefficients between measured and predicted uncertainties are shown in Table 1.

Table 1. Correlation coefficients between measured and predicted uncertainties for samples with different ^{238}U/^{232}Th ratios

^{238}U/^{232}Th	Correlation coefficient							
	^{226}Ra	^{238}U	^{235}U	^{40}K	^{232}Th	^{134}Cs	^{137}Cs	^{7}Be
0.63	0.9770	0.9568	0.9914	0.9765	0.9353	0.9612	0.9915	0.9291
0.77	0.9604	0.9540	0.9758	0.9625	0.9326	0.9463	0.9743	0.9221
1.72	0.9248	0.9234	0.9421	0.9229	0.9152	0.9150	0.9382	0.9050

What is more important is the fact that the difference between actual and ANN predicted uncertainties are not critically influenced by the absolute values of uncertainties in detecting particular radionuclides (Fig. 2) as the sign of differences in uncertainties is not a function of the measuring time.

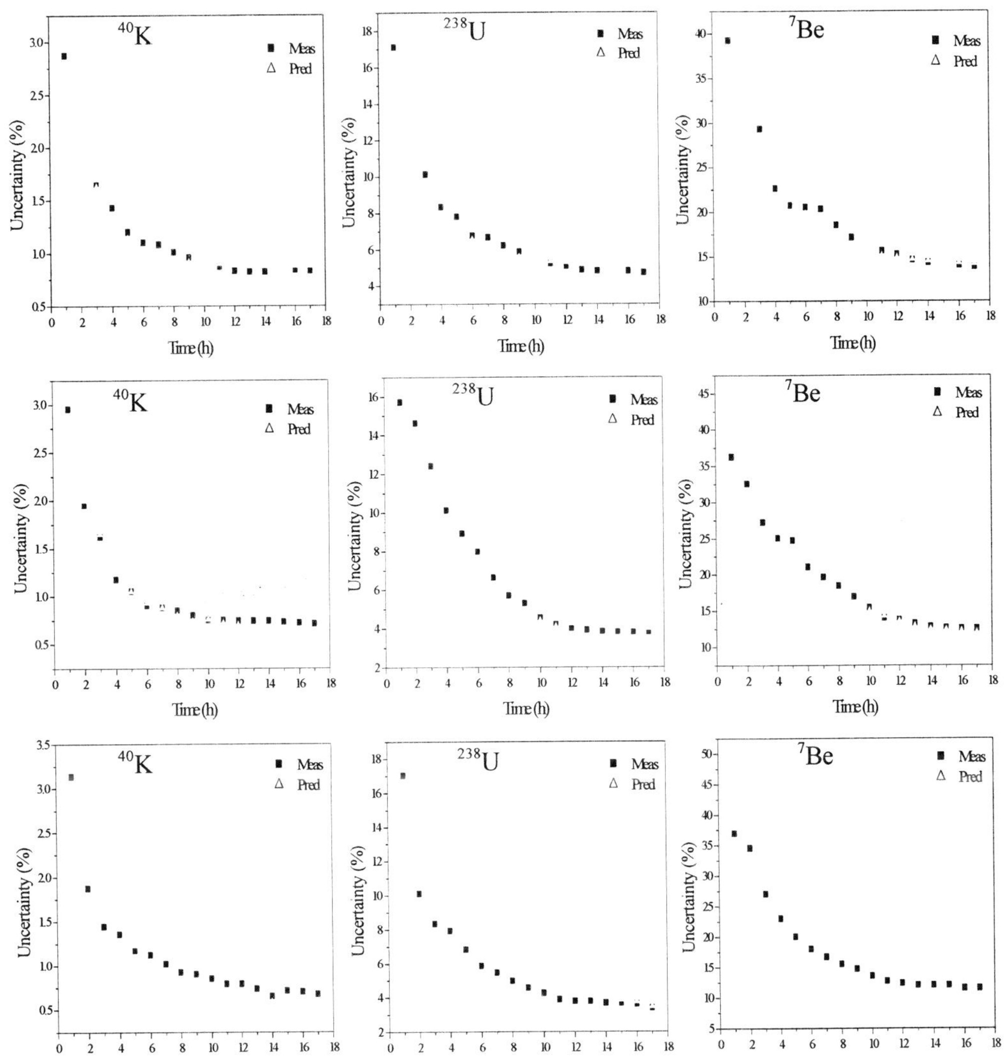

Figure 2. Experimental and predicted uncertainties of measurement of activities of three radionuclides (^{40}K, ^{238}U and ^{7}Be) in samples with different ^{238}U/^{232}Th ratios (top row: 0.63; middle row: 0.77; and bottom row: 1.72) as a function of measurement time.

Namely, irrespective of whether we are examining radionuclides with low (^{40}K) or high (^{7}B) uncertainties, the two curves intermesh in a random fashion (Fig. 2; top row). Also, the correlation coefficient appears not to be related to the measuring uncertainties. The same pattern was found in samples with different ^{238}U/^{232}Th ratios (Fig. 2; middle and bottom row), but the correlation coefficients for all radionuclides were lower than those established for the $^{238}U/^{232}Th$ ratio used for training the network. The differences were negligible for the sample with the $^{238}U/^{232}Th$ ratio of 0.77, but slightly increased (approximately by 4%) for the sample with a ratio of 1.72. These differences were not significant, i.e., once ANN was trained it could be successfully used for uncertainty modeling regardless of the content of radionuclides in soil samples ($^{238}U/^{232}Th$ ratio).

3.1.2. Optimization of Peak-to-Background Ratio

The peak-to-background ratio (PBR) is of considerable importance for the optimization of detection limits in gamma spectrometric measurements (Semkow et al., 2002). It is well known that detection limits can be improved by optimizing the measurement time. Because PBRs of many radionuclides must be optimized in one spectrum, this task is often complex and time-consuming. Simplex optimization of ANN model parameters (learning rate ( ), momentum ( ), number of epochs (E), and the number of nodes in the hidden layer (N)) was performed in order to design an ANN for the prediction of PBR as a function of measurement time in gamma-ray spectrometric analysis of environmental soil samples (Dragović and Onjia, 2005)[2].

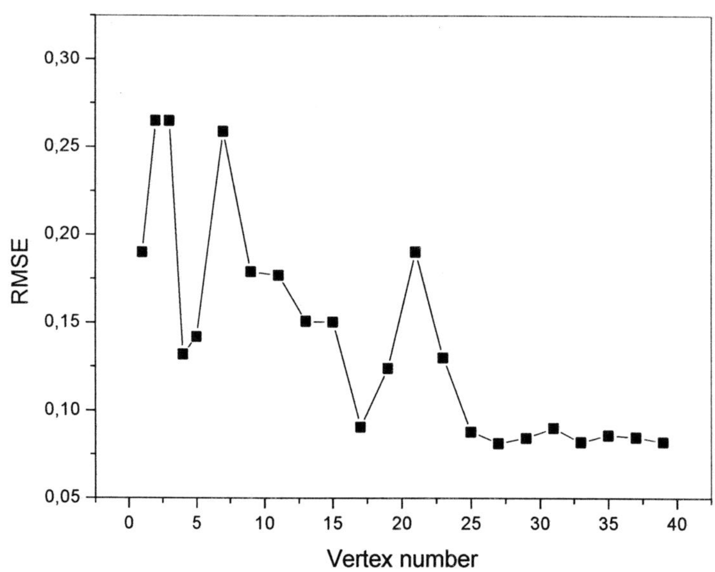

Figure 3. Evaluation of RMSE with trials defined by the simplex method.

A three-layer feed-forward ANN with a back propagation algorithm was used to model the PBR in gamma spectrometric analysis as a function of measurement time. Delta-bar-delta adaptive learning rate method in which each weight has its own learning rate was used. Modified simplex algorithm was implemented according to the description given by Nelder and Mead (Nelder and Mead, 1965). Simplex calculations were made using a laboratory-

[2] Figures in this section are reprinted from Dragović, S.; Onjia, A. *Appl. Radiat. Isot.* 2005, 63, 363-366, with kind permission from Elsevier.

written Mathcad program. As four parameters were to be varied simultaneously, the initial simplex consisting of the first five vertices was established by the first design matrix of coordinates. The initial simplex was comprised of , ranging from 0 to 100, , from 0 to 1, E from 1 to 100000 and N from 2 to 20. The simplex was moved in the direction given by the rules of the variable-size simplex algorithm. The data obtained from the simplex procedure are presented in Fig. 3. The vertex number 29 gave the optimum values for the parameters, however, due to a relatively large noise in the system the optimization lasted up to vertex 39 and the parameter values were averaged (excluding outliers) for the last 10 vertices. The averaged parameter values for the minimum RMSE are: = 20, = 0, E = 45000 and N = 5.

The optimized ANN model was used to predict PBRs of radionuclide activities for 14 measurement times, which were not used for training the network. The predictive power of the ANN model was examined by comparing the experimental PBRs with the predicted ones.

The relative accuracy was calculated from experimental and ANN predicted PBR values:

$$\% \text{ Relative accuracy} = \frac{PBR_{pred} - PBR_{exp}}{PBR_{exp}} \cdot 100 \tag{22}$$

where PBR_{exp} and PBR_{pred} is the PBR determined experimentally and by ANN, respectively (Fig. 4).

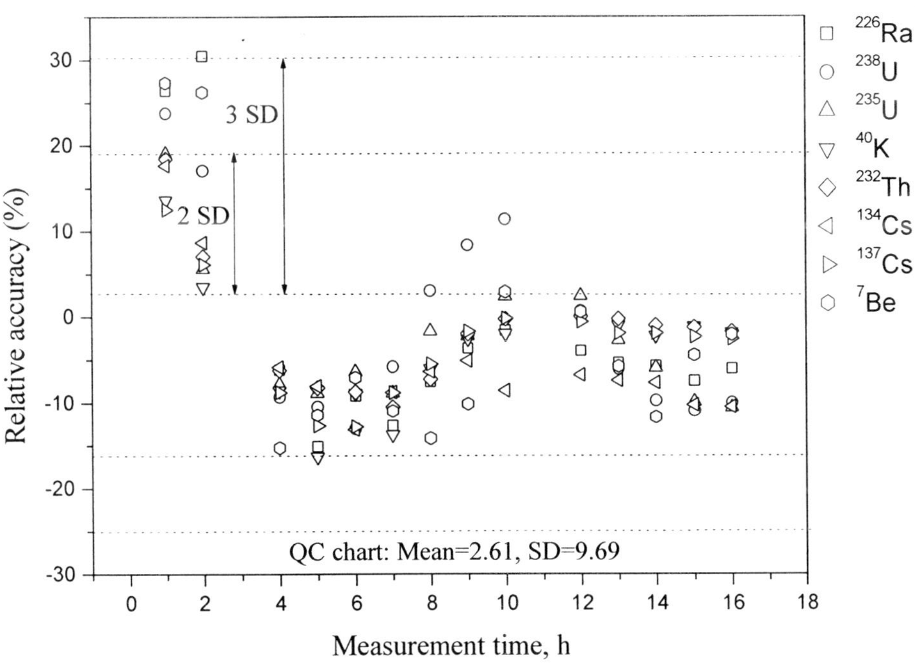

Figure 4. Relative accuracy as a degree of agreement between the measured and the predicted peak-to background ratios for eight radionuclides.

It can be seen that the mean value of the relative accuracy and the standard deviation (SD) for all studied radionuclides is 2.61 and 9.69%, respectively. The predicted and experimental PBR values for the studied radionuclides agreed to within ± 19.4 % of expanded uncertainty (level of confidence 95%) with an average bias of 2.61%, which was considered adequate.

These ANN performances are valid for the case studied, though ANN prediction of PBR may be more or less accurate, depending on the particular radionuclide/matrix system, as well as on the experimental inputs for ANN training. The aim of the study was to avoid traditional optimization approach by using trial-and-error method. On the other hand, so called "hard" modeling using numerical methods based on interpolation performs equally well as ANN. We also made a PBR prediction by using quadratic model. The average correlation coefficient for eight radionuclides between experimental data and data obtained by ANN prediction was 0.9939, while the average correlation coefficient between experimental data and data obtained by simple interpolation was 0.9905.

The advantage of ANN over numerical methods is that the optimization can be done very fast, no mathematical form of the relationship between the input and the output data is necessary, the analyst can be unfamiliar with the system. These advantages are particularly significant in gamma-ray spectrometric analysis of a large number of environmental samples.

3.1.3. Optimization of Minimum Detectable Activity

A performance criterion of gamma-ray spectrometry is the minimum detectable activity (A_D), defined qualitatively as the smallest amount of radionuclide that can be reliably determined, given the prevailing conditions of the particular spectral measurement. According to Currie (Currie, 1995), the activity value for which the false negative error is β, giving such a critical activity value A_C that the probability of the estimated quantity exceeding A_C is not greater than α if analyte is absent ($A_C = 0$, null hypothesis; α is the false positive error), is an accurate quantitative definition. In another formulation, the A_D is the true net activity for which the probability that the estimated value does not exceed A_C is β, where A_C is the minimum significant value of an estimated activity for which the probability that A_C exceeds the null net activity is α. Values $\alpha=\beta=0.05$ (recommended within the analytical chemistry community) have been adopted in gamma-ray spectrometry for the computation of A_D (95% confidence level).

The widely used simplex algorithm proposed by Nelder and Mead (Nelder and Mead, 1965) was applied to optimize the ANN parameters, because of the algorithm's simplicity and efficiency. The ANN model optimized by the simplex method was employed to predict the minimum detectable activity of radionuclides in environmental soil samples of different compositions (Dragović et al., 2006)[3].

Minimum detectable activities for eight radionuclides from three different measurement times were used to train the network.

As four parameters (learning rate, momentum, number of nodes in the hidden layer, and number of epochs) are varied simultaneously, the initial simplex consisting of the first five vertices was established by the first design matrix of coordinates. The network parameters ranged from 1 to 100 for the learning rate, from 0 to 1 for the momentum, from 1 to 100000 for the number of epochs, and from 2 to 20 for the number of nodes in the hidden layer. The

[3] Figures and tables in this section are reprinted from Dragović, S.; Onjia, A.; Bačić, G. *Nucl. Instr. Meth. Phys. Res. A* 2006, 564, 308-314, with kind permission from Elsevier.

simplex was designed to include relatively large changes in factor levels (α=1-100, E=1-100000) in order to cover a large portion of factor space. This strategy, contrary to that when only small variations in factor levels are used, usually produces a more rapid convergence to the optimum through contraction than would be achieved if only small variations in factor levels were used and the simplex is required to expand.

The simplex was moved in the direction given by the rules of the variable-sized simplex algorithm. The system response was evaluated with RMSE (Fig. 5).

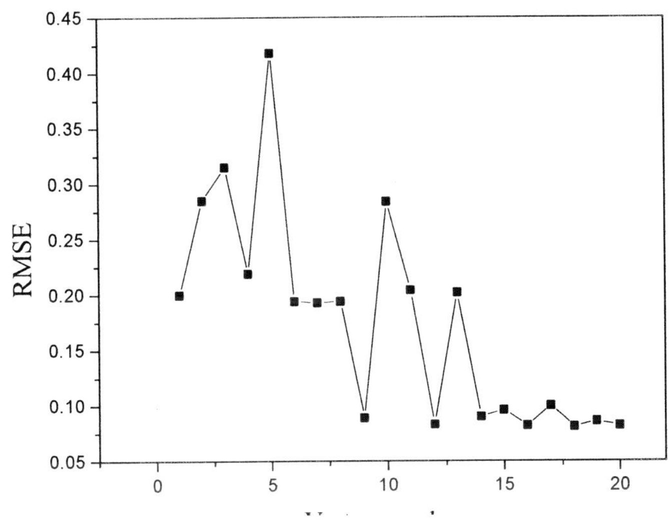

Figure 5. The evaluation of RMSE with trials defined by the simplex method.

The average parameter values for the minimum RMSE were: α = 15.76; μ = 0.27; N = 19 and E = 37000.

The architecture of the optimized ANN consisted of an input layer with one node representing the time of measurement (t), an output layer with eight nodes representing the A_D of the investigated radionuclides (^{226}Ra, ^{238}U, ^{235}U, ^{40}K, ^{232}Th, ^{134}Cs, ^{137}Cs and ^{7}Be), and a hidden layer with the number of nodes to be optimized. There were also biases connected to nodes in the hidden and the output layer via modifiable weighted connections, responsible for accommodation of non-zero offsets in the data.

Experimental versus predicted A_D scatter plots obtained using the ANN model optimized by the simplex method is shown in Fig. 6.

These plots with correlation coefficients from 0.9811 for ^{232}Th to 0.9995 for ^{40}K demonstrate the ability of the ANN model to predict the A_D of radionuclides in the analyzed soil samples. Neither the differences between the experimental and the ANN-predicted A_D values or the correlation coefficients were influenced by the absolute values of the A_D of the investigated radionuclides (illustrated in Fig. 6). Therefore, in spite of significant differences in the A_D for ^{226}Ra and ^{137}Cs, the correlation coefficients between the experimental and predicted A_D values were similar (0.9875 and 0.9890, respectively).

The performance of the ANN model developed here was evaluated by its prediction of the A_D of radionuclides in soil samples of different compositions, specifically the uranium-to-thorium ($^{238}U/^{232}Th$) ratios. The correlation coefficients between measured and predicted A_D for eight radionuclides in samples are given in Table 2.

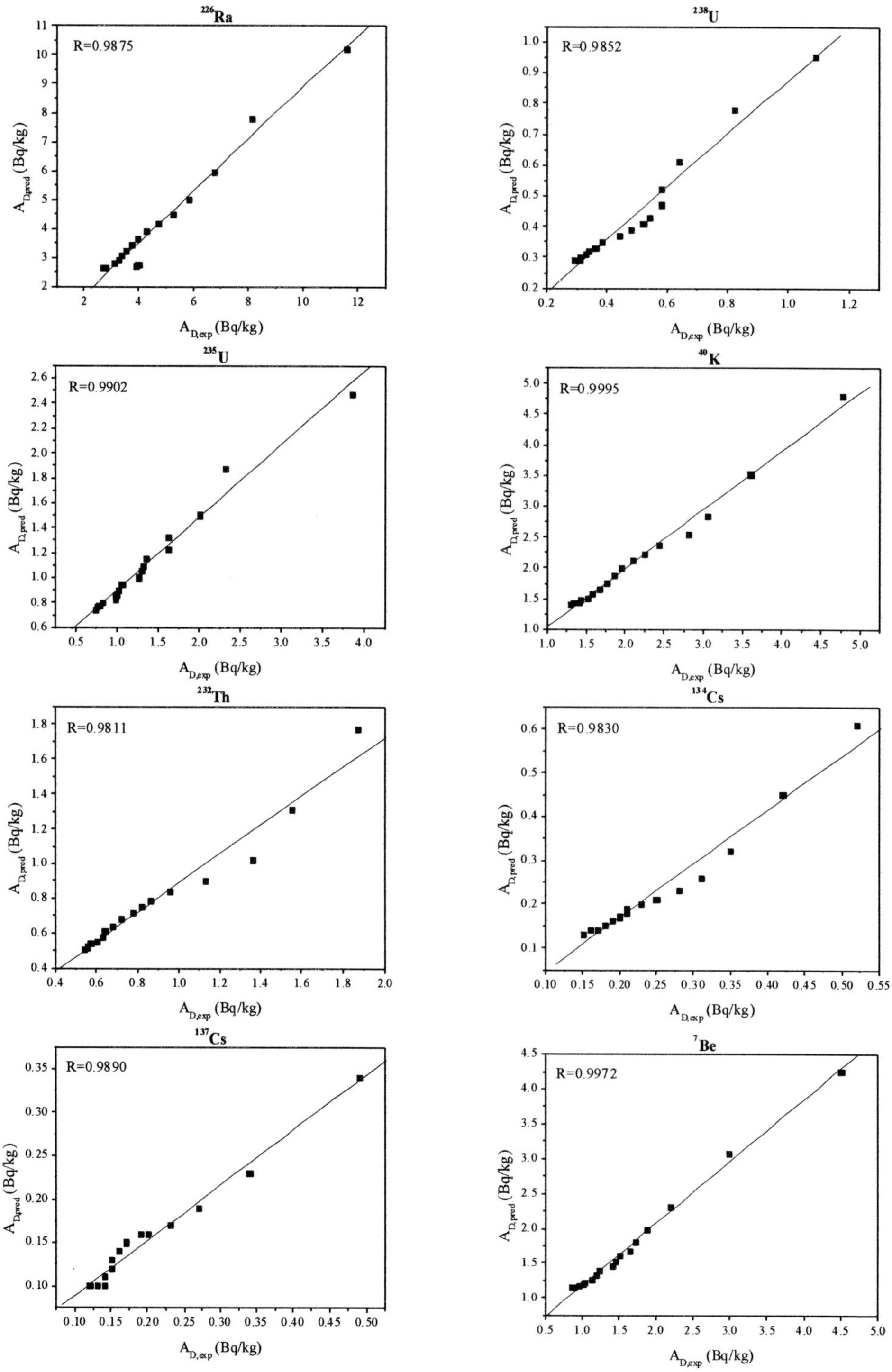

Figure 6. Experimental versus predicted A_D values using the simplex optimized ANN model.

Table 2. Correlation coefficients between measured and ANN predicted A_D values for soil samples with different $^{238}U/^{232}Th$ ratios

$^{238}U/^{232}Th$	Correlation coefficient							
	^{226}Ra	^{238}U	^{235}U	^{40}K	^{232}Th	^{134}Cs	^{134}Cs	^{7}Be
0.43	0.9875	0.9852	0.9902	0.9995	0.9811	0.9830	0.9890	0.9972
0.51	0.9954	0.9891	0.9820	0.9847	0.9880	0.9905	0.9912	0.9900
0.85	0.9801	0.9793	0.9890	0.9784	0.9836	0.9808	0.9879	0.9842
0.92	0.9790	0.9863	0.9712	0.9710	0.9600	0.9825	0.9840	0.9803
1.14	0.9907	0.9949	0.9902	0.9871	0.9517	0.9623	0.9616	0.9828

The obtained results confirmed the predictive ability of the optimized model, regardless of the content of radionuclides in the soil samples. In addition, a successful prediction of the A_D by the ANN model was possible, even when the network was trained with a small quantity of experimental data. The accuracy of prediction obtained was acceptable for most applications in gamma-ray spectrometry. ANN analysis is quite flexible as regards to the amount and form of the training (experimental) data, indicating that the use of more informal experimental designs than those in statistical approaches is possible. Since the accuracy of an ANN model is strongly influenced by the quality of training data, it is important that the training data display as many variations as possible and completely represent the entire set of data to be analyzed.

The ANN model outperformed the quadratic model of simple interpolation (our unpublished data). Compared with hard models, ANNs indicated higher flexibility and were less complex in many other applications (Zeng et al., 2001; Despagne and Massart, 1998). The advantage of an ANN over numerical methods based on interpolation is that the optimization can be performed very quickly, no mathematical form of the relationship between the input and the output data is necessary, and the analyst can be unfamiliar with the system. This is particularly significant in gamma-ray spectrometric analysis of a large number of environmental samples.

When systematic approaches are lacking, the optimization of ANNs is performed by "trial and error" or by changing one parameter at a time whilst keeping the others constant. Often, it is necessary to find a reliable, accurate and robust optimization approach to select the optimal ANN parameters. As demonstrated in many applications, simplex optimization satisfies all the above-mentioned points (Üstün et al., 2005; Jiang et al., 2003; Sartoros et al., 1998).

The simplex optimization procedure was developed as a relatively fast alternative to the "trial and error" method usually applied to optimize the parameters of a particular ANN model. The performance of ANN model depends sensitively on the learning rate (α) and momentum (μ) parameters. A good choice of these parameters is very essential for training success and speed. Adjusting these parameters by "trial and error" can be very difficult and might take a long time for more complicated tasks. As the size of the network increases, the network becomes more computationally intensive, and so the time required to train the network grows exponentially and the learning process becomes unacceptably slow (Kim et al., 2000). In our work, the trial and error method show several fold slower convergence than the simplex method (Dragović, 2005) due to a large number of unsuccessful trials. The method was not efficient because of entrapment into local minima where it could not escape in practical time. The principal advantage of the simplex method is its ability to avoid this entrapment and hence to search for a global optimum. The simplex method identifies the optimum response with fewer trials compared with non-systematic approaches or one-variable-at-a-time methods. Clearly the use of the simplex optimization method is a valuable tool when faced with a number of time-consuming applications.

3.2. Classification of Samples in Environmental Studies

3.2.1. Classification of Soil Samples According to Their Geographic Origin Using Different Pattern Recognition Techniques

It is of considerable interest to be able to determine the geographic origin of soils analytically. The geographic origin can be recognized with minimum effort if the relevant constituents are analyzed and the results included in data analysis algorithms. A number of analytical techniques and methods may be used in the classification of soil samples of different taxonomy and/or geographic origin (Eswaran et al., 2002). In the last few years, concentration profiles of chemical substances in conjunction with pattern recognition methods have been widely employed to classify soils (Ramadan et al., 2001; Antonić et al., 2003; McBratney et al., 2003; Slavković et al., 2004).

The specific problem was to discriminate soil samples from Serbia and Montenegro according to geographic origin (Dragović and Onjia, 2006; Dragović and Onjia, 2007; Dragović and Onjia, 2007a)[4]. Soil samples were collected from different regions and analyzed by gamma-ray spectrometry. The samples were then classified according to their origin based on their radionuclide (^{226}Ra, ^{238}U, ^{235}U, ^{40}K, ^{134}Cs, ^{137}Cs, ^{232}Th and ^{7}Be) activities. Natural environmental radioactivity depends on the geological and geographic conditions, and appears at different levels in the soils of each different geological region (UNSCEAR, 2000). The specific levels of terrestrial environmental radiation are related to the geological composition of each lithologically separated area, and to the content of natural radionuclides in rocks from which the soils originate in each area. Geologically, the territory of Serbia and Montenegro includes a great number of rock complexes (magmatic, sedimentary and metamorphic rocks) which are markedly different in respect to age, genesis, mineral content and petrochemical and geochemical characteristics. Outstanding differences in natural radioactivity of soils can exist in relation to their geological origin (Dimitrijević, 1995; Dragović et al., 2006a; Dragović et al., 2006b). Therefore, the set of natural radionuclides is used to differentiate investigated areas based on geology. The concentration of cesium isotopes are influenced by altitudes of sampling areas. Ecosystems on higher altitudes are predisposed to receive more fallout and therefore higher concentrations of these radionuclides (Howard et al., 1991; Dragović et al., 2004). The concentrations of beryllium on the ground are higher in areas of high rainfall. Since precipitation is generally higher in upland regions, an increase in concentration of beryllium with altitude is to be expected (Salisbury and

[4] Figures and tables in this section are reprinted from Dragović, S.; Onjia, A. *J. Environ. Radioact.* 2006, 89, 150-158 and Dragović, S.; Onjia, A. *Appl. Radiat. Isot.* 2007, 65, 218-224, with kind permission from Elsevier.

Cartwright, 2005). Due to these reasons, cesium isotopes and beryllium were used in our work to differentiate areas according to their susceptibility to fallout.

Samples of undisturbed soils were taken from fifteen locations in Serbia and Montenegro during 2003. Descriptive statistics of radionuclide activities in the soil samples are given in Table 3.

Table 3. Descriptive statistics and results of Shapiro-Wilk's test of normality of radionuclide activities (Bq/kg) in soil samples collected from Serbia and Montenegro

Descriptive statistics								
Parameter	^{226}Ra	^{238}U	^{235}U	^{40}K	^{232}Th	^{134}Cs	^{137}Cs	^{7}Be
Mean activity	30.8	29.7	1.36	567	40.7	0.1	48.3	1.79
Uncertainty	1.4	1.2	0.2	9.8	1.1	0.1	0.4	0.5
Median	29.9	30.2	1.38	593	39.7	0.05	42.9	1.50
Mode	32.7	15.4	1.14	728	45.3	0.02	31.2	0.70
Standard deviation	9.14	9.42	0.462	163.6	13.54	0.074	26.19	1.045
Skewness	0.42	0.44	0.53	-0.17	1.20	0.91	0.44	0.66
Kurtosis	-0.45	-0.33	0.22	-0.73	2.18	-0.56	-0.39	-0.91
Range	41.3	38.8	2.10	648	65.1	0.24	106.8	3.40
Minimum	13.6	14.6	0.51	271	18.3	0.02	5.25	0.54
Maximum	54.9	53.4	2.61	919	83.4	0.26	112	3.94
Shapiro-Wilk's statistics								
R	0.989	0.986	0.986	0.978	0.940	0.947	0.986	0.946
P-value	0.083	0.043	0.037	<0.010	<0.010	<0.010	0.035	<0.010

Radionuclide activities in these samples varied by factors of up to 3-5 for ^{226}Ra, ^{238}U, ^{235}U, ^{40}K and ^{232}Th, 7 for ^{7}Be, 13 for ^{134}Cs, and 21 for ^{137}Cs. The standard deviation was the greatest for ^{40}K and the smallest for ^{134}Cs. The arithmetic mean and the standard deviation of the activities for all samples and locations were used to describe the central tendency and variation of the data. Grubb's test, an important procedure for removing outliers from the dataset, was applied prior to classification (Grubb, 1969). This resulted in no outliers detected in the dataset.

Shapiro-Wilk's test (significance level α was 0.05) (Shapiro, 1965) for normality of activity distribution within each radionuclide was applied first (Table 1) and revealed normal distribution of the data.

The feasibility of using pattern recognition techniques to predict soil origin based on its radionuclide content was examined. Five common pattern recognition techniques, principal component analysis (PCA), linear discriminant analysis (LDA), *k*-nearest neighbours (KNN), soft independent modelling of class analogy (SIMCA) and artificial neural network (ANN), were employed to classify soil samples based on their radionuclide contents determined by gamma-ray spectrometry. The data set consisted of a 103×8 matrix, in which rows represented the soil samples analyzed (103 objects), and columns the specific activities of eight radionuclides determined by gamma-ray spectrometry (8 variables). Each soil sample was represented in the 8th-dimensional space by a data vector which is an assembly of the 8 features. Data vectors belonging to the same category (sample origin) were analyzed by PCA, LDA, KNN, SIMCA and ANN. Multivariate data analyses were performed by means of the software packages: SPSS 10.0 (SPSS Inc., 1999), PLS_Toolbox 2.1 (Eigenvector Research Inc., 1998), and QwikNet 2.23 (QwikNet Version 2.23, 1999).

The number of significant principal components (PCs), which represent linear combinations of the original radionuclide activities, was selected on the basis of the Kaiser criterion (Kaiser, 1960). This criterion retains only PCs with eigenvalues that exceed one. The PCA, applied to the given dataset, has shown a differentiation between the soil samples according to their geographic origin (Fig. 7).

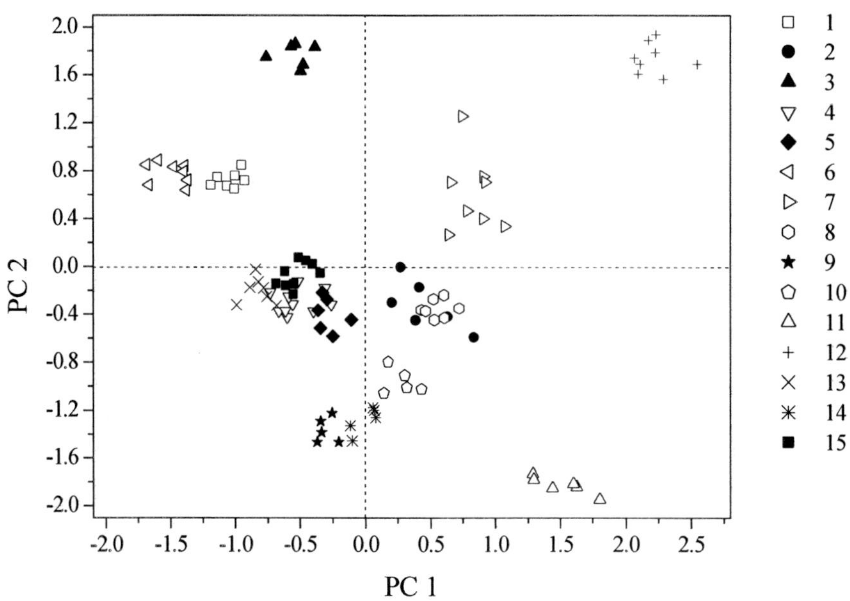

Figure 7. Score plot of PC1 and PC2 illustrating the differentiation of soil samples according to their geographic origin.

Quantitative estimation of the classification efficiency was obtained from the classification matrix (Table 4), which shows the number of samples in each class and percentage of correctly classified samples.

Table 4. Classification matrix

Class	**1**	**2**	**3**	**4**	**5**	**6**	**7**	**8**	**9**	**10**	**11**	**12**	**13**	**14**	**15**
No. of samples	7	6	6	10	6	8	8	7	5	5	6	8	7	5	9
Corr. class. (%)	100	66.7	100	50.0	50.0	100	100	85.7	100	100	100	100	71.4	100	66.7

An overall 86.0% correct classification was achieved. Samples from nine locations were accurately classified (100%). Some samples from the other locations were misclassified, i.e. the classification rate ranged from 50 to 85.7%. Some samples were incorrectly assigned as they fell on the borders between their own class and another one.

The results obtained in this study were in good correlation with those obtained in many other studies dealing with soil classification. Thus, PCA was shown to be an efficient tool for soil classification based on the content of trace elements (Slavković et al., 2004) or polycyclic aromatic hydrocarbons (Golobočanin et al., 2004). The application of PCA in forensic classification of soils of different origin (Thanasoulias et al., 2002) gave a classification efficiency of 85%, which is close to the efficiency obtained in this study.

LDA produced a recognition rate of 95.1%; however, prediction ability was less satisfactory (82.8%). Fig. 8 shows the separation of soil samples by two discriminant functions.

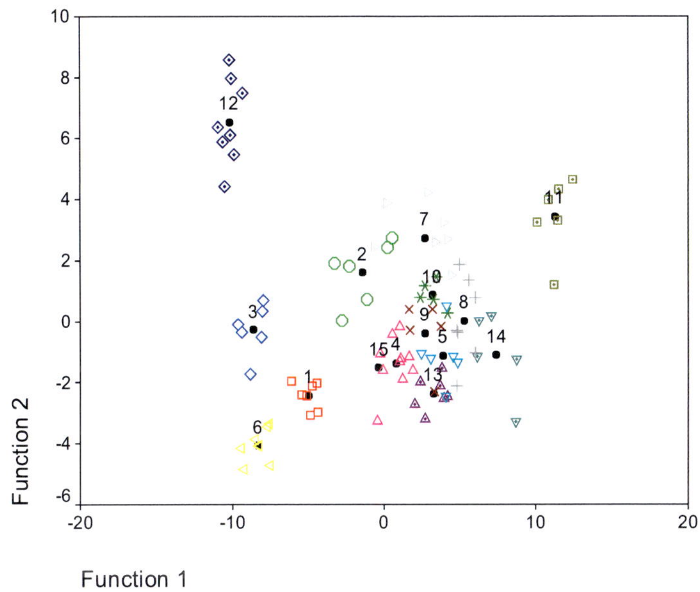

Figure 8. Projection of soil samples from 15 different regions in the space formed by the discriminant functions after LDA.

The highest rate of recognition achieved by KNN classifier was 86.4% and that of prediction was 88.6%. The number of neighbours (*k*) assayed were 1,3,5,7 and 9. For k value of 1, four samples from test set were misclassified, for *k* values of 3 and 5, seven samples were misclassified, for k values of 7 and 9, eleven and thirteen samples were misclassified, respectively. In Table 5 recognition and prediction abilities of KNN classifier for different *k* values are presented.

Table 5. Recognition and prediction ability of KNN for different number of neighbours (*k*) used for classification

k	Recognition ability (%)	Prediction ability (%)
1	86.4	88.6
3	84.5	80.0
5	77.7	80.0
7	70.9	68.6
9	68.0	62.9

The results obtained by SIMCA were very poor (Table 6), giving only 60.0% of correct assignation for samples from test set containing autoscaled data. Meancentering of the data resulted in prediction ability of only 42.8%.

Table 6. Recognition and prediction ability of SIMCA method obtained with two types of preprocessing

Type of preprocessing	Recognition ability (%)	Prediction ability (%)
Meancenter	86.4	42.8
Autoscale	86.4	60.0

The ANN classifier used was architectured by an input layer consisting of the radionuclide activities in the soil samples, a hidden layer, and an output layer composed of regions the samples were collected from. The strategy used for the classification of soils was the so-called one-of-many encoding (Di Natale et al., 2001). The output of the network was a multi-dimensional vector and each vectorial dimension class was assigned to a class. In the network training file, the class membership of a single data item was coded in a numerical format, assigning 1 to the belonging class and 0 to all the others. In the network testing file, the membership of an input data set was assigned to the class with the greatest network output. Different combinations of the learning rate and momentum were used to find the best configuration for model optimization, in order to allow the solution to escape the local minima of the error function. For this particular case, the learning rate (η) and momentum (μ) were set to 0.1. The number of epochs was also optimized. The optimum number of learning epochs was found to be 2500. The number of nodes in the hidden layer is the most important parameter of the network architecture which strongly influences its prediction accuracy. To optimize this parameter, ANNs with different number of nodes in hidden layer were trained.

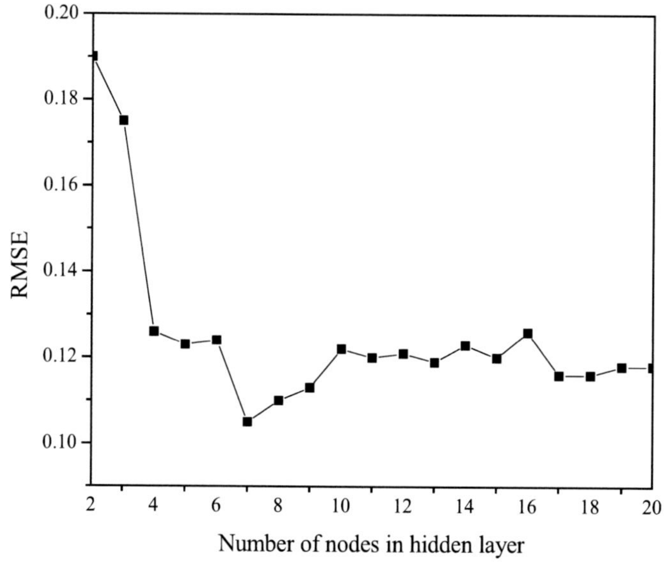

Figure 9. The RMSE for different number of nodes in hidden layer of ANN model.

Each topology was repeated five times to avoid random initialization of weights. Through the above process, it was found that the optimum number of hidden layer nodes was 7 (Fig. 9).

Table 7 shows recognition and prediction abilities afforded with each multivariate technique. The recognition ability of these methods ranged from 86.4 to 100% and the prediction ability from 60.0 to 92.1%. In this application, ANN method over performed the classification capabilities of other techniques. All soil samples presented to the network were correctly classified, i.e. the recognition ability was 100% and the prediction ability was 92.1% (Table 7).

Table 7. Recognition and prediction ability of PCA, LDA, KNN, SIMCA, and ANN to classify soil samples

Method	Recognition ability (%)	Prediction ability (%)
PCA	100.0	86.0
LDA	95.1	82.8
KNN (K=1; inverse squared Euclidean distance)	86.4	88.6
SIMCA (normal range; α=0.05)	86.4	60.0
ANN (8×7×15; η=0.1; μ=0.1; sigmoidal transfer function)	100.0	92.1

Comparing the ANN and KNN methods, it is evident that the most significant advantage of KNN classifier is that no training is required for classification. However, this investigation has proved that the generalization capability of the neural network is an unavoidable characteristic for this application and that this is an overriding advantage over the KNN method and its simplicity. In addition, KNN method is very sensitive to gross inequalities in the number of objects in each class, while ANN does not impose any condition on the data structure.

The advantages of neural network methodology, particularly the ability to extract hidden features from the input signals, can be expected to be particularly useful when the data sets with very complicated correlation structures are to be analyzed or when data sets contain series of many highly inter correlated variables. Neural network are able to reproduce the training data in much more detail than other statistical techniques, which is particularly important if the specific problem under consideration requires a more detailed solution, which the usual statistical approach based on means and variances may be unable to provide. The main advantage of neural network-based decision models lies in their adaptive behavior.

3.2.2. Implementation of Neural Networks for Classification of Moss and Lichen Samples on the Basis of Gamma-Ray Spectrometric Analysis

Due to their peculiar anatomical, morphological and physiological characteristics (high sorption capacity, water exchange) mosses and lichens are useful bioindicators of environmental pollution particularly for a variety of metals and radionuclides of both natural and artificial origin (Nifontova, 1995; Figuera et al., 2002; Kirchner and Daillant, 2002). Mosses and lichens exhibit several advantages as monitors of atmospheric deposition due to their lack of a protective cuticle and thickened cell walls, numerous cell wall constituents with negatively charged groups, mineral nutrition mainly from wet and dry deposition. They are distributed over large geographical areas, therefore allowing comparison of the pollutant content in ecosystems from diverse regions. The accumulating capacity of mosses and lichens are several-fold higher than those observed in vascular plants growing in the same habitat. The morphology of mosses and lichens does not vary throughout the seasons, thus accumulation of pollutants can occur throughout the year. These species usually have considerable longevity, which assists their use as long-term integrators of atmospheric deposition (Pucket, 1988; Tyler, 1990). The degree of their sensitivity to environmental pollution varies between species and with respect to the type of pollutant (Zechmeister et al., 2003; Dragović et al., 2004).

ANNs were employed to classify mosses and lichens according to their geographical origin based on their content of radionuclides of different origin (members of the natural uranium and thorium decay chains, cesium isotopes originating from the Chernobyl powerplant accident and cosmogenic beryllium) determined by gamma-ray spectrometry. The feasibility of five different ANN training algorithms were examined and their classification efficiencies were compared (Dragović et al., 2007)[5]. The optimised ANN was able to predict the origin of moss and lichen samples which were not used for the network training.

Samples of mosses and lichens (n=120) were collected from twelve geographic regions of Serbia and Montenegro: Kopaonik, Zlatibor, Durmitor, Biogradska gora, Bjelasica, Bukulja, Kosmaj, Avala, Belgrade, Fruška Gora, Vršac and Vranje during the period from 2000 to 2003.

The descriptive statistics of radionuclide activities (Bq/kg) in moss and lichen samples are presented in Table 8. Radionuclide activities varied by a factor of 6-12 for ^{226}Ra, ^{238}U, ^{235}U and ^{7}Be, 19 for ^{40}K, 25 for ^{134}Cs and 1328 for ^{137}Cs.

Table 8. Descriptive statistics of radionuclide activities (Bq/kg) in the moss and lichen samples

Parameter	^{226}Ra	^{238}U	^{235}U	^{40}K	^{232}Th	^{134}Cs	^{137}Cs	^{7}Be
Mean	49.5	31.9	1.59	272	34.5	0.09	1575	109
Median	42.9	30.8	1.48	215	33.5	0.05	734	98.6
Mode	92.3	17.6	0.70	92.0	28.0	0.01	470	115
Standard deviation	25.5	12.4	0.67	172.4	13.92	0.008	2200	59.5
Skewness	0.37	0.80	0.58	1.02	1.57	0.47	2.20	0.55
Kurtosis	-1.18	0.38	-0.27	0.60	4.34	-1.42	4.11	-0.71
Range	87.2	58.3	3.00	758	78.0	0.24	9293	230
Minimum	11.2	14.8	0.40	42.0	16.0	0.01	7.00	20.0
Maximum	98.4	73.1	3.40	800	94.0	0.25	9300	250

Since the distribution parameters (skewness and kurtosis) indicated lack of the normal distribution, the data were logarithmically transformed.

[5] Figures and tables in this section are reprinted from Dragović, S.; Onjia, A.; Dragović, R.; Bačić, G. *Environ. Monit. Assess.* 2007, 130, 245-253, with kind permission of Springer Sciences and Business Media.

The architecture of the ANN consisted of an input layer representing the radionuclide activities in moss and lichen samples, a hidden layer and an output layer composed of regions the samples were collected from. By way of the optimization process, the optimum values of the network parameters were obtained. The values which gave the minimum RMSE were 0.1 for the learning rate and momentum, 5 for the number of nodes in the hidden layer and 3000 for the number of epochs (Figure 10).

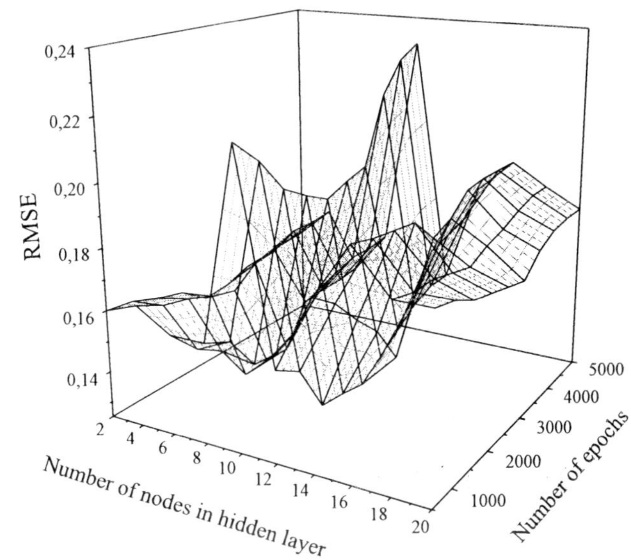

Figure 10. The influence of the number of epochs and the number of hidden layer nodes on the RMSE.

Table 9. The classification matrix obtained by an online back-propagation randomized network. The values are averaged for different training/testing set combinations

Actual class	Output of classifier (%)											
	1	2	3	4	5	6	7	8	9	10	11	12
1	92	0	0	0	0	0	6	0	0	0	2	0
2	0	91	0	0	0	5	0	0	0	0	0	4
3	0	0	90	0	0	0	0	10	0	0	0	0
4	0	0	0	88	6	0	0	0	0	0	6	0
5	4	0	0	0	93	0	0	0	3	0	0	0
6	0	1	1	0	0	95	0	0	3	0	0	0
7	0	0	0	0	0	0	91	0	0	0	0	9
8	0	0	0	2	0	0	0	89	0	9	0	0
9	0	0	9	0	1	0	0	0	90	0	0	0
10	4	0	0	0	0	0	0	1	4	91	0	0
11	0	8	0	0	0	0	0	0	0	0	92	0
12	0	0	0	10	0	0	3	0	0	0	0	87

ANN model validation was performed by presenting the test set to the optimized network. The predictive power was assessed through the classification matrix, in which the classes predicted by the ANN classifier were compared to the actual classes (Table 9). An ideal classifier would produce a classification matrix in which all diagonal elements were 100%, while all off-diagonal elements were 0%. The matrix in Table 9 is not ideal, but the origin of most samples was correctly classified. The average classification rate, defined as the ratio of the number of correctly identified samples to that of all test samples (the average of diagonal elements in the classification matrix), was 90.7 ± 4.8%.

When classification was based only on the activities of natural radionuclides, a lower average classification rate 86.2 ± 4.5% was obtained (Table 10).

Table 10. The classification matrix obtained by an online back-propagation randomized network with the exclusion of ^{134}Cs and ^{137}Cs activities

Actual class	Output of classifier (%)											
	1	2	3	4	5	6	7	8	9	10	11	12
1	87	0	0	0	0	8	0	0	0	5	0	0
2	0	84	0	0	10	0	0	0	0	0	0	6
3	0	0	94	0	0	0	0	3	0	0	0	3
4	0	0	0	81	0	0	10	0	0	0	9	0
5	0	0	0	5	83	0	0	0	12	0	0	0
6	7	0	0	0	0	86	0	0	0	0	7	0
7	0	6	0	0	0	0	82	0	0	10	0	2
8	0	0	0	0	0	0	0	96	1	0	3	0
9	0	10	0	0	3	0	0	0	87	0	0	0
10	0	0	0	14	0	0	0	1	0	85	0	0
11	6	0	6	0	0	0	8	0	0	0	80	0
12	0	0	0	0	4	6	0	0	0	0	1	89

There are already a number of well known and widely used chemometric methods that provide spatial estimates and/or predictions according to data obtained in environmental monitoring studies. Geostatistical methods offer a way of describing the spatial continuity that is an essential feature of many natural phenomena. In recent years, the number of applications using geostatistical models applied to environmental sciences has increased resulting in successful results on modelling soil, water and air pollution (Pereira et al., 1997). Factor-analytical methods have already been applied in environmental studies in which mosses and lichens were used as biomonitors. Principal component analysis applied to the concentrations of 26 elements in the moss *H. splendens* revealed the main origin and distribution of metal pollutants in Norway (Berg et al., 1995). Monte-Carlo assisted factor analysis has been applied to a lichen data set obtained from a monitoring survey on a national scale in The Netherlands for source apportionment purposes (Kuik and Wolterbeek, 1995). This analysis was found to be suitable for simplifying large and complex environmental data sets in such a way that it may create a limited number of factors, each representing a cluster of interrelated variables within a data set (for example, geographically dependent correlations between element concentrations).

The RMSE value for predictions on the test set and the average classification rate of 90.7% indicate a good generalization ability of the ANN model optimised. The ability of the ANN to extract hidden features from the input signals was found to be particularly useful for this kind of monitoring when the data sets with complex correlation structures had to be analyzed or when data sets contained series of many highly inter-correlated variables.

The lower classification rate based on activities of only natural radionuclides could be explained by the fact that the majority of nutrient uptake by mosses and lichens is from the atmosphere and clear correlation exists between the content of radionuclides in radioactive fallouts (^{134}Cs and ^{137}Cs) and in these species. The misidentification of some moss and lichen samples indicates that some other characteristics of these species also influence the pattern of activities of the eight commonly detected radionuclides. Radionuclide accumulation by mosses and lichens is also influenced by (micro)climate, pH, rainfall, chemical forms of radionuclides, tolerance, age, growth rate and chemical and/or morphological properties of the species. Interpretation of unexplained variability could be attributable to these meteorological and ecological characteristics. Previous studies have shown that the use of a large number of variables could decrease the classification capability of ANN (Suchacz and Wesołowski, 2006). Therefore, before including these parameters into ANN modelling data sets for classification of moss and lichen species, data reduction methods (principal component analysis) must be applied to eliminate factors with insignificant contribution to the total variance of the data set.

Conclusion

Five commonly used pattern recognition methods (PCA, LDA, kNN, SIMCA, ANN) were tested in different applications for purposes of monitoring of radioactive contamination of the environment. The accuracy of ANN models in optimization of important parameters of gamma-ray measurements: measuring uncertainty, peak-to-background ration and minimum detectable activity were examined. The results obtained indicate that the neural network approach is a suitable alternative to the traditional approach in environmental radioactivity studies. Moreover, because output results for a given set of input data are available in a very short time (a few seconds), ANN has considerable potential for solving time-consuming problems in systematic analysis of environmental samples.

Pattern recognition methods were successfully applied for classification of environmental samples, frequently used in monitoring programmes according to geographic origin. In all cases, prediction ability was between 83 and 92%. Some soil samples were incorrectly assigned to one or two wrong classes that were not always geographic neighbors. This misidentification of the soil origin could indicate that other soil characteristics also influence the pattern of activities of detected radionuclides. The misidentification of some moss and lichen samples indicates that meteorological conditions of the investigated areas, as well as ecological characteristics, also influence the pattern of activities of the eight commonly detected radionuclides. Further research is needed in this specialized aspect of biomonitoring to fully quantify the influence of these parameters on the classification accuracy of classification model.

Acknowledgments

The authors gratefully acknowledge financial support from the Ministry of Science of the Republic of Serbia (Contract No. 142039).

References

Antonić, O.; Pernar, N.; Jelaska, S.D. *Ecol. Model.* 2003, 170, 363-371.

Berg, T.; Røyset, O.; Steinnes, E.; Vadset, M. *Environ. Pollut.* 1995, 88, 67-77.

Brereton, R.G. *Data Analysis for the Laboratory and Chemical Plant*; John Wiley & Sons; Ltd: West Sussex, 2003.

Coomans, D.; Jonckheer, M.; Massart, D.; Broeckeart, I.; Blockx, P. *Anal. Chim. Acta* 1978. 103, 409-415.

Currie, L.A. *Pure Appl. Chem.* 1995, 67, 1699-1723.

De Maesschalck, R.; Candolfi, A.; Massart, D.L.; Heuerding, S. *Chemom. Intell. Lab. Sys.* 1999, 47, 65-77.

Despagne, F.; Massart, D.L. *Analyst* 1998, 123, 157R-178R.

Di Natale, C.; Macagnano, A.; Martinelli, E.; Proietti, E.; Paolesse, R.; Castellari, L.; Campani, S.; D'Amico, A. *Sens. Act. B* 2001, 77, 561-566.

Dimitrijević, M., *Geology of Yugoslavia*; Geoinstitute, Belgrade (in Serbian), 1995.

Dragović, S. *Application of neural networks in gamma-ray spectrometric analysis of soil and plant samples*; PhD thesis, Faculty of Physical Chemistry, University of Belgrade, 2005.

Dragović, S.; Janković, Lj.; Onjia, A. *Radiat. Prot. Dosim.* 2006a, 121, 297-302.

Dragović, S.; Janković, Lj.; Onjia, A.; Bačić, G. *Radiat. Meas.* 2006b, 41, 611-616.

Dragović, S.; Nedić, O.; Stanković, S.; Bačić, G. *J. Environ. Radioact.* 2004, 77, 381-388.

Dragović, S.; Onjia, A. *Appl. Radiat. Isot.* 2005, 63, 363-366.

Dragović, S.; Onjia, A. *Appl. Radiat. Isot.* 2007, 65, 218-224.

Dragović, S.; Onjia, A. *J. Environ. Radioact.* 2006, 89, 150-158.

Dragović, S.; Onjia, A. *Russ. J. Phys. Chem. A* 2007a, 81, 1477-1481.

Dragović, S.; Onjia, A.; Bačić, G. *Nucl. Instr. Meth. Phys. Res. A* 2006, 564, 308-314.

Dragović, S.; Onjia, A.; Dragović, R.; Bačić, G. *Environ. Monit. Assess.* 2007, 130, 245-253.

Dragović, S.; Onjia, A.; Stanković, S.; Aničin, I.; Bačić, G. *Nucl. Instr. Meth. Phys. Res. A* 2005, 540, 455-463.

Eigenvector Research, Inc., PLS_Toolbox 2.1, Manson, WA, 1998.

Einax, J.W.; Soldt, U. *Chemom. Intell. Lab. Sys.* 1999, 46, 79-91.

Einax, J.W.; Zwanzinger, H.W.; Geiss, S. *Chemometrics in Environmental Chemistry*; VCH, Weinham, 1997.

Eswaran, H.; Rice, T.; Shrens, R.; Stewart, B.A. *Soil Classification: A Global Desk Reference*; CRC Press, 2002.

Figuera, R.; Sérgio, C; Sousa, A.J. *Environ. Pollut.* 2002, 118, 153-163.

Frean, M.; Robins, A. *Catastrophic forgetting: a review and an analysis of the pseudo-rehearsal solution;* Report AIM-36-97-2, Artificial Intelligence Laboratory, Computer and information Science, University of Otago, Dunedin, New Zealand, 1977.

Fukunaga, K. *Introduction to statistical pattern classification*; Academic Press, San Diego, 1990.

Golobočanin, D.D.; Škrbić, B.D.; Miljević, N. *Chemom. Intell. Lab. Syst.* 2004, 72, 219-223.

Golub, G.H.; Van Loan, C.F. *Matrix computations*; The Johns Hopkins University Press, Baltimore, 1996.

Grubb, F. *Technometr.* 1969, 11, 1-21.

Hastie, T.; Tibshirani. R.; Friedman, J. *The elements of statistical learning: data mining, inference, and prediction*, Springer, 2001.

Hopke, P.K. *Anal. Chim. Acta* 2003, 500, 365-377.

Hopke, P.K. *Atmosph. Environ.* 1988, 22, 1777-1792.

Howard, B.J.; Beresford, N.A.; Hove, K. *Health Phys.* 1991, 61, 715-725.

Jain, A.K.; Duin, R.P.W.; Mao, J. *IEEE Trans. Pattern Anal. Mach. Intell.* 2000, 22, 4-37.

Jiang, J.H.; Liang, Y.Z.; Ozaki, Y. *Chemom. Intell. Lab. Syst.* 2003, 65, 51-65.

Joliffe, I.T. *Principal Component Analysis*; Springer, New York, 1986.

Jorgensen, S.E. *Ecol. Model.* 1976, 2, 147-165.

Kaiser, H.F. *Edu. Physiol. Meas.* 1960, 20, 141-151.

Kanevski, M.; Arutyunyan, R.; Bolshov, L.; Demyanov, V.; Maignan, M. *Geoinform* 1996, 7, 5-11.

Kanevski, M.; Demyanov, V.; Maignan, M. *Conference ICANN 1997,* Lausanne, 1997.

Kangas, L.J.; Troyer, G.L.; Keller, P.E.; Hashem, S.; Kouzes, R.T. *IEEE NPSS Conference Proceedings* 94CH35762, November 1994, p. 48.

Keller, P.E.; Kangas, L.J.; Hashem, S.; Kouzes, R.T.; Troyer, G.L. *Proceedings of the 1995 Workshop on Environmental and Energy Applications of Neural networks*, World Scientific, Singapore, 1995.

Keller, P.E.; Kouzes, R.T. *IEEE NPSS Conference Proceedings 94CH35762*, November 1994, p. 341.

Keller, P.E.; Kouzes, R.T.; Kangas, L.J.; *IEEE Electro/94 International Conference*, Boston, 1994, p. 378.

Kim, J.; Mowat, A.; Poole, P.; Kasabov, N. *Chemom. Intell. Lab. Syst.* 2000, 51, 201-216.

Kirchner, G; Daillant, O. *Environ. Pollut.* 2002, 120, 145-150.

Kuik, P.; Wolterbeek, H.Th. *Water Air Soil Poll.* 1995, 84, 323-346.

Maier, H.R.; Dandy, G.C. *Environ. Model. Softw.* 1998, 13, 193-209.

Mardiyanto, M.P.; Uritani, A.; Sakai, H.; Kawarabayashi, J.; Iguchi, T. *Nucl. Instr. Meth. Phys. Res. A* 2000, 441, 483-488.

Massart, D.; Vandeginste, B.; Deming, S.; Michotte, Y.; Kaufman, L. *Chemometrics: A Textbook*; Elsevier, New York, US, 1998.

Massart, D.L.; Vandeginste, B.G.M.; Buydens, L.M.C.; de Jong, S.; Lewi, P.J.; Smeyers-Verbeke, J. *Handbook of Chemometrics and Qualimetrics*; Elsevier Science, Amsterdam, 1998a.

McBratney, A.B.; Mendonça Santos, M.L.; Minasny, B. *Geoderma* 2003, 117, 3-52.

Mitchell, T.M. *Machine Learning*; McGraw-Hill International Editions, 1997.

Nelder, J.A.; Mead, R. *Comput. J.* 1965, 7, 308-313.

Nifontova, M. *Sci. Total Environ.* 1995, 160-161, 749-752.

Olmos, P.; Diaz, J.C.; Perez, J.M.; Aguago, P.; Gomez, P.; Rodellar, V. *IEEE Trans. Nucl. Sci.* 1994, 41, 637-641.

Olmos, P.; Diaz, J.C.; Perez, J.M.; Gomez, P.; Rodellar, V.; Aguayo, P.; Bru, A.; Garcia Belmonte, G.; de Pablos, J.L. *IEEE Trans. Nucl. Sci.* 1991, 38, 971-975.

ORTEC, Gamma Vision 32, Gamma-Ray Spectrum Analysis and MCA Emulation, Version 5.3, Oak Ridge, USA, 2001.

Özesmi, S.L.; Özesmi, U.; Tan, C.O. *Methodological issues in building, training and testing artificial neural networks*; Cornell University preprint archive arXiv: q-bio.PE/0510017, 2005.

Penza, M.; Cassano, G.; Tortorella, F. *Meas. Sci. Technol.* 2002, 13, 846-858.

Pereira, H.G.; Ribeiro, L.; Soares, A.; Patinha, P.; Pereira, M.J.; Dias, M. In *Geostatistics Wollongong*; E.Y. Baafi; A. Schofield (Eds.); Kluwer Academic Pub., Dordrecht, 2, 1997, pp. 951-960.

Pilato, V.; Tola, F.; Martinez, J.M.; Huver, M. *Nucl. Instr. Meth. A* 1999, 422, 423-427.

Pucket, K.J. In *Lichens, Bryophytes and Air Quality*; Bibliotheca Lichenologica 30, 231-267. T.H. Nash III and V. Wirth (Eds.), J. Cramer, Berlin-Stutgart, 1988.

QwikNet Version 2.23, Craig Jensen, Redmond, 1999, http://qwiknet.home.comcast.net/.

Ramadan, Z.; Song, X.H.; Hopke, P.K.; Johnson, M.J.; Scow, K.M. *Anal. Chim. Acta* 2001, 446, 233-244.

Reifman, J.; Vitela, J.E. *Nucl. Techn.* 1994, 106, 225-241.

Ross, G.G. *J. Theor. Biol.* 1976, 56, 381-399.

Rumelhart, D.; Hinton, G.; William, R. In *Proceedings Parallel Distributed Processing. Foundations*; Rumelhart, D., McClelland, J. (Eds.), MIT Press, Cambridge, 1986, vol. 1.

Sakai, H.; Uritani, A.; Takenaka, Y.; Mori, C.; Iguchi, T. *Nucl. Instr. Meth. A* 1999, 421, pp. 316-321.

Salisbury, R.T.; Cartwright, J. *J. Environ. Radioact.* 2005, 78, 353-361.

Sartoros, C.; Goltz, D.M.; Salin, E.D. *Appl. Spectrosc.* 1998, 52, 643-648.

Semkow, T.M.; Parekh, P.P.; Schwenker, C.D.; Khan, A.J.; Bari, A.; Colaresi, J.F.; Tench, O.K.; David, G.; Guryn, W. *Appl. Radiat. Isot.* 2002, 57, 213-223.

Shapiro, S.S.; Wilk, M.B. *Biometr.* 1965, 52, 591-611.

Sharma, A.; Paliwal, K.K.; Onwubolu, G.C. *Patt. Recogn.* 2006, 39, 1215-1229.

Slavković, L.; Škrbić, B.; Miljević, N.; Onjia, A. *Environ. Chem. Lett.* 2004, 2, 105-108.

Sloof, J.E.; Wolterbeek, H.Th. *Water, Air, Soil Poll.* 1991, 57-58, 785-795.

SPSS 10.0 for Windows, http://www.spss.com/.

Stratis, J. A.; Tsakovski, S.; Simeonov, V.; Zacharaidis, G.; Sawidis, T. *Toxicol. Environ. Chem.* 1998, 00, 1-10.

Stratis, J. A.; Zacharaidis, G.; Simeonov, V.; Mandjukov, P.; Sawidis, T. *Toxicol. Environ. Chem.* 1996, 54, 29-37.

Suchacz, B.; Wesołowski, M. *Talanta* 2006, 69, 37-42.

Tan, C.O.; Bekligou, M. *Ecol. Model.* 2005, 183, 425-434.

Tan, C.O.; Özesmi, U.; Beklioglu, M.; Per, E.; Kurt, B. *Ecol. Inform.* 2006, 1, 195-211.

Thanasoulias, N.C.; Piliouris, E.T.; Kotti, M.S.E.; Ermiridis, N.P. *For. Sci. Intern.* 2002, 130, 73-82.

Tsakovski, S.; Simeon, V.; Sawidis, T.; Zachariadis, G.; Stratis, J. *Toxicol. Environ. Chem.* 1999, 69, 287-294.

Tyler, G. *Bot. J. Linn. Soc.* 1990, 104, 231-253.

UNSCEAR. *Sources and effects of ionizing radiation;* Report to General Assembly, with Scientific Annexes, United Nations, New York, 2000.

Üstün, B.; Melssen, W.J.; Oudenhuijyen, M.; Buydens, L.M.C. *Anal. Chim. Acta* 2005, 1-2, 292-305.

Vigneron, V.; Morel, J.; Lepy, M.C.; Martinez, J.M. *Nucl. Instr. Meth. A* 1996, 369, 642-647.

Walley, W.J.; O'Connor, M.A. *Ecol. Model.* 2001, 146, 219-230.

Werbos, P.J. *Beyond regression: new tools for predictive and analysis in the behavioral sciences*; Ph.D. thesis, Harvard University, Cambridge, MA, 1974.

Wold, S. *Pattern Recog.* 1976, 8, 127-139.

Wu, W.; Massart, D.L. *Chemom. Intell. Lab. Sys.* 1996, 35, 127-135.

Yiftah, S. *Neural networks - potential application in the nuclear industry: preliminary comments and reflections. Use of Expert Systems in Nuclear Safety*; International Atomic Energy Agency, IAEA-TECDOC-542, 1988, 177-182.

Yoshida, E.; Shizuma, K.; Endo, S.; Oka, T. *Nucl. Instr. Meth. Phys. Res. A* 2002, 484, 557-563.

Zechmeister, H.G.; Grodzińska, K.; Szarek-Łukaszewska, G. In *Bioindicators and Biomonitors*; Markert, B.A., Breure, A.M.; Zechmeister (Eds.), Elsevier, London, 2003, pp. 329-375.

Zeng, Y.B.; Xu, H.P.; Liu, H.T.; Wang, K.T.; Chen, X.G.; Hu, Z.D.; Fan, B.T. *Talanta* 2001, 54, 603-609.

In: Pattern Recognition in Nanoscience...
Editor: Diane M. Perez, pp. 159-180

ISBN: 1-60021-715-X

Chapter 6

USING MULTIVARIATE STATISTICS FOR PATTERN RECOGNITION IN ARCHAEOLOGY

Grant S. McCall*

Department of Anthropology, 114 MacBride Hall,
University of Iowa, Iowa City, IA 52242

Abstract

Pattern recognition is of primary importance to the field of archaeology, given that recognizing patterns is the key to both understanding problems with our current beliefs about the past and the generation of new and better ideas to resolve them. This chapter discusses the use of multivariate statistics for data reduction and pattern recognition in large archaeological datasets. Most archaeological research is at least partially quantitative, but most of this research has traditionally focused on hypothesis testing and the calculation of probability statistics. While statistical hypothesis testing is an important aspect of scientific approaches to archaeological research, the field has largely overlooked the use of non-probability based statistical approaches to pattern recognition. The chapter begins with a brief discussion of the philosophy of science and the role of archaeological pattern recognition in the construction of accurate generalizations about the archaeological record based on material remains, the construction of hypotheses based on these generalizations, and the critical testing of hypotheses based on new and better understandings of archaeological patterning. This chapter then outlines the use of multivariate statistical approaches including factor analysis, principal components analysis, and multiple regression, which are powerful and under-utilized approaches to large archaeological datasets. The chapter illustrates the use of these techniques by discussing several important case study analyses of some major categories of materials and situations commonly encountered by archaeologists.

Introduction

There are many good discussions of the use of multivariate statistical techniques in archaeology (Baxter 1994, 2003; Shennan 1997). Based on these, it would be easy to for an

* E-mail address: grant-mccall@uiowa.edu
DO NOT CITE IN ANY CONTEXT WITHOUT PERMISSION OF AUTHOR

archaeological researcher with a cursory knowledge of statistics to learn how to use widely available statistical software packages to calculate useful multivariate statistics. What is lacking from these discussions is a clear explanation of why any sane archaeologist would want to bother going through this imposing process. Let us begin by facing this fact: Most practicing archaeologists are not keen to spend their scarce free time reading statistics texts in order to learn more about approaches that are mathematically difficult and not obviously necessary to ongoing research. Furthermore, the ability to easily add new statistical techniques to our *repertoire* left most of us many years and brain cells ago. From a rhetorical perspective, this places the onus on me to explain how and why multivariate statistical techniques could be important tools for archaeologists in a readable way. This is not a reference piece, as there are many excellent specimens already available (e.g. Baxter 1994, 2003; Shennan 1997; and many others). Rather, this is intended to be a gateway for archaeologists interested in applying multivariate statistical techniques to their own research problems. This chapter assumes a certain level of statistical knowledge, but uses plain language and humor as ways of appealing to readers for whom statistics is not their favorite topic of light reading. The point of my discussion is that multivariate statistical techniques are powerful and under-exploited tools for archaeological pattern recognition, and that these approaches are more accessible than is generally thought.

In both the past and the present, dynamic human behavior results in *patterns* of material remains distributed in three-dimensional space. Archaeology is fundamentally based on the premise that patterns of archaeological remains were structured by activities of living people, and therefore that archaeological patterns may be used to make inferences about patterns of human behavior in the past. Because of the relationship between human behavior and the patterns of archaeological remains it produces, the importance of pattern recognition for archaeology should be self-evident. The key to making inferences about the past rests directly on accurate pattern recognition. Archaeological patterns are the direct sources of the ideas we hold concerning prehistory. Likewise, we discover problems with our views on the past through the discovery of contradictory patterning, and this stimulates the formation of new ideas. In short, to paraphrase Binford (1981), pattern recognition is the key to noticing our areas of ignorance about the past, and for resolving them.

It is also the case that most modern archaeological research is at least partially quantitative on some level. The process of fieldwork involves recording the location of important archaeological objects and features in three-dimensional space. Laboratory work involves counting various categories of objects, and counting or measuring certain features of objects. For this reason, archaeological patterns are usually recognized and described through quantitative methods. Statistical approaches for pattern recognition should be of utmost importance to archaeological studies. Surprisingly, the use of pattern recognition statistics in archaeology is relatively uncommon, and most statistical techniques that archaeologists use are aimed at the assessment of probability and hypothesis testing. This strong focus on probability, or what I might tongue-in-cheek call "p-value envy," has an important place in deductive archaeological science, but it is not the only productive use of statistics.

It is important to point out that archaeological patterns are inherently multivariate phenomena. Clearly, the explanation of human behavior itself, which is one of the main goals of scientific archaeology, relies on the ability to understand the simultaneous operation of many environmental and social variables. Archaeologically, many forces affect the formation of sites and the characteristics of recovered assemblages. For this reason, multivariate statistical techniques are important ways of getting at this complexity in terms of both human behavior and natural forces of site formation. This chapter proposes that multivariate statistical techniques for pattern recognition are powerful tools for quantitatively studying the archaeological record.

The chapter begins by exploring pattern recognition in archaeology from the perspective of philosophy of science, and reviews the place of pattern recognition in terms of both the construction and testing of models. Next, the chapter reviews several important multivariate statistical techniques, including factor analysis (FA), principal components analysis (PCA), correspondence analysis (CA) and multiple regression / partial correlation. The chapter illustrates the potential uses of these statistical techniques through a number of cases studies concerning types of materials commonly encountered by archaeologists. These include (1) Binford's (1981) study of animal bones from Olduvai, Tanzania, (2) Bernardini's (2005) study of ceramic decoration and rock art in the American Southwest, (3) Wilson's (1998) study of rock art elements in Oceania, (4) my own work (McCall 2006) on the stone tool industries of Klasies River Mouth, South Africa, and (5) O'Connell's (1987) ethnoarchaeological study of the Australian Alyawara. Hopefully, these case studies will provide useful guides for researchers dealing with similar kinds of problems in their own research.

Beyond "P-Value Envy": Archaeology, Pattern Recognition, and the Philosophy of Science

The potential for a discussion of multivariate statistical techniques to be incredibly boring is already high, but when combined with talk of philosophy of science, it becomes immeasurable. For this reason, I will try to keep a sense of humor in this section, and remain as brief as possible. It is also important to point out that I would never begin this discussion if it were not absolutely necessary to the main points of the chapter in terms of using statistics to make inferences about the past, and to build theory concerning the organization of human behavior over time.

The formation of the New Archaeology during 1960s led to a heightened awareness of the philosophy of science and a focus on archaeology as deductive science. As Flannery (1986: 511) quips, this involved a focus on Popper (mainly 1959) and Hempel (mainly 1965), and other philosophers of science "many of them named Karl." It is easy to be critical of this movement today, given changes in both the field of archaeology and the emergence of philosophical positions based in post-modernism. But it is impossible to deny the importance of these fundamental critiques in terms of rebuking a then largely discovery-driven field, and in showing the clear dangers of strict generalizing empiricism. Binford (1968) especially saw the problems with empirical generalizations about the past. He saw that such research resulted in the cumulative generation of archaeological descriptions of patterns and scenarios to account for them, but no clear way to recognize bad ideas or develop new ones. The significant contribution of deductive approaches based on falsification have been well-described elsewhere (e.g. Trigger 1989), so I will spare further discussion here.

Pattern recognition is important to any scientific enterprise, and this is definitely true for archaeology. Deductive scientific archaeology, of the kind advocated during the rise of the

New Archaeology, depends on the recognition of patterns for two purposes: (1) The generation of models of past human behavior based on generalizations stemming from patterned characteristics of the archaeological record, and (2) the deductive testing of these models using patterned archaeological evidence. In short, archaeologists use patterns of archaeological evidence to both build models and to critically test them. No discussion of the philosophy of science would be complete without a flowchart, and mine is present in Figure 1. This flowchart shows the importance of pattern recognition to both the inductive and deductive aspects of scientific archaeological research.

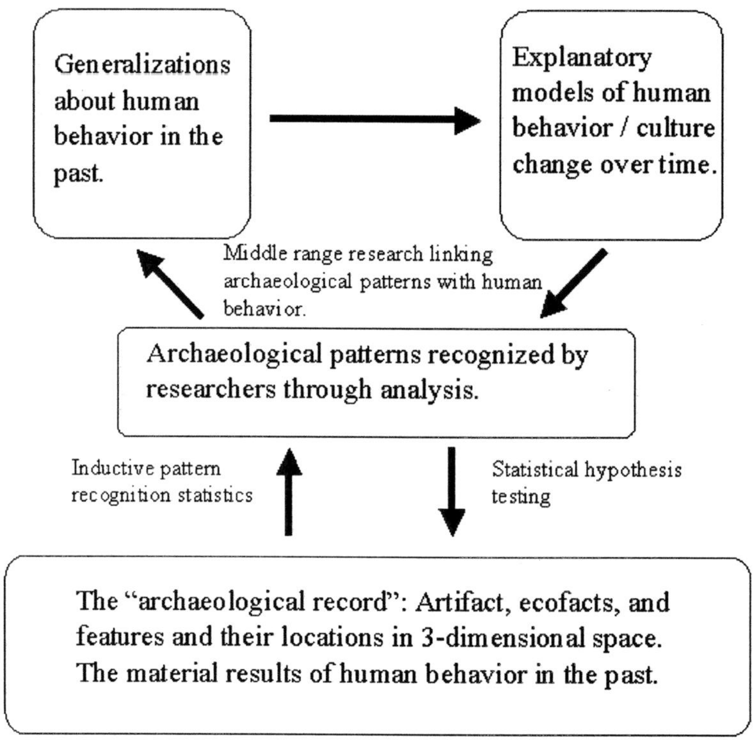

Figure 1. Flowchart representing the processes of scientific archaeological research.

The archaeology of the 1970s and 1980s focused on the deductive hypothesis-testing aspects of scientific archaeological research. In terms of statistical methods, this led to the dominance of probability assessment approaches to quantitative data. This emphasis on statistical hypothesis testing and probability assessment still dominates modern quantitative archaeological research. While most archaeological studies do not go to the fullest lengths in terms of framing null and alternative hypotheses, test implications, and other deductive scientific trappings, a high frequency do report some kind of statistical test resulting in a statement of probability, or a p-value. Such statistical tests can be both parametric and non-parametric (although frequently ignorant of the assumptions behind these terms), and include chi-square tests, t tests, F tests, ANOVA tests, correlation, regression, and many other derivatives of these that are too numerous to usefully list. What is frequently missed in the use of these statistical techniques is that their actual utility is in assessing the relationship between a sample and the population which it represents. These tests essentially answer the question, What is the probability that some observed characteristic of a sample could have resulted from chance alone, rather than actually representing some real characteristic of the population from which it was selected?

There is no doubt that these issues of probability and sampling are very important to archaeology, which is entirely dependent upon sampling. However, there is frequently a sense in archaeological studies that statistical analyses resulting in a p-value greater than 0.05 (the traditional cutoff for rejecting a null hypothesis) have somehow come up with "bad results." Likewise, studies with statistical tests resulting in very low p-values are frequently taken as powerful evidence for a strong statistical relationship, and therefore a high level of potency in terms of the original idea being tested. Ignoring the problematic conflation of probability and strength of association, which is almost never reported in archaeological studies, this situation creates an irrational lust after low p-values, or what I might call "p-value envy." While we clearly need statistical methods for assessing probability and hypothesis testing, this is not the only effective quantitative learning strategy available for archaeological research. It is incumbent on the researcher to (1) know the actual assumptions behind the statistical tests being used, (2) know what question the hypothesis testing is actually addressing, and (3) to be smart about using statistical tests and their results to explore patterns in the archaeological record and to test models of their causes.

If most statistical approaches used in archaeology are not pattern recognition tools, then how are patterns actually recognized? The answer to this question usually involves general descriptive statistics and the eyeballing of relationships. For example, imagine two Neolithic villages are excavated, and one is found to have a higher mean number of ground stone axes per household than the other. The investigating researchers then perform a two-sample t test, resulting in a p-value of less than 0.05. The fact that one village has more ground stone axes per household than the other is then used to address some deeper theoretical issue concerning the organization of human behavior in the past. The pattern was recognized when the researchers found that one village had a higher mean number of axes per household-- a basic descriptive statistic. The t test did not help discover that pattern, but rather simply tested whether the pattern could be explained as the result of chance in the process of sampling. Recognizing the pattern through descriptive statistics and then testing it using probability statistics were separate tasks, but both necessary for establishing that the observed pattern was real and appropriate for examining important theoretical questions.

Often, situations occur where there are too many variables for descriptive statistics to be effective tools for pattern recognition. For instance, returning to the Neolithic village example, imagine that the researcher wants to compare not only the frequency of ground stone axes, but also forty other categories of object. Or imagine that the researchers want to compare a number of measurements (mass, length, width, thickness, raw material, color, etc.) on forty different categories of object. Also, imagine if it were twenty villages instead of only two. This could be done by eyeballing descriptive statistics, but it would be an incredibly time-consuming and painstaking process involving thousands of head-to-head comparisons. Furthermore, such comparisons have difficulty in examining how different variables relate to one another, at least without adding another hugely labor-intensive layer of descriptive statistical analysis. For example, the frequency of axes might covary with the frequency of sickle handles, and this might be an important archaeological pattern for addressing research questions. Finding this pattern would only be possible if the frequency of every category of object was compared with the frequency of every other category of object at every site. With forty categories of object and twenty sites, it would take literally thousands of head-to-head comparisons to explore every possibility. While archaeologists may follow hunches in exploring patterns using descriptive statistics, it is impossible to deal with such complex quantitative situations in a systematic way without more sophisticated multivariate statistical techniques.

Unfortunately, the last scenario described above is a fairly common situation faced by archaeologists. The result is that most archaeologists will do as many head-to-head comparisons of descriptive statistics as human ability, time, and budget will allow, and also pick a specific (but deliberately small) set of variables based largely on hunches to explore on an *ad hoc* basis. The complexity of this situation puts a premium on *data reduction* and the use of more sophisticated multivariate pattern recognition techniques in dealing with larger datasets.

Returning to issues of the philosophy of science, multivariate statistical techniques are important for identifying archaeological patterns that serve as the basis for the inference of human behavior in the past. As Binford (1981) proposes, the main strategy archaeologists have for learning about the past relies on establishing the relationship between dynamic, living human behavior and the static material arrangements it produces, or the archaeological record. Binford calls this process "middle range" research, emphasizing the importance of ethnoarchaeological observations of modern peoples as a way of learning the processes that produced the archaeological record in the past. Multivariate pattern recognition statistics are often used during the inductive phases of research, serving as the basis for generalizations about the archaeological record, and the source of scenarios of human behavior in the past. Hypothesis-testing statistics are usually used during deductive phases of research, and they are based on the premise of falsifying existing models (Popper 1959). They require both *a priori* knowledge of what important variables actually are, and specific existing hypotheses to be tested with know quantitative parameters.

It is important to point out that pattern recognition statistics, while less capable of calculating probability and testing statistical hypotheses, are equally able to call into question inferences of past human behavior, and therefore falsify models explaining behavior. When archaeological patterns are found that contradict fundamental understandings of human behavior in the past, this highlights important areas of scientific ignorance. This is integral to the process of deductive science; it leads to the abandonment of older, problematic models and the generation of newer, more elegant ones. In short, hypothesis-testing statistics are most useful during the deductive phases of research when specific variables have already been identified as crucial and there are defined models being tested through a process of falsification. In contrast, multivariate pattern recognition statistics are most useful in reducing data, identifying what variables are important, and systemically recognizing patterning between variables in terms of covariation.

The other reason for using multivariate statistics as opposed to bivariate (or univariate) approaches is that both human behavior and the formation of the archaeological record itself are inherently multivariate phenomena. Bivariate models are really only effective in dealing with phenomena where only one major independent variable explains only one other dependent variable. For example, when Newton examined acceleration due to gravity, this was possible using bivariate models, since gravity was the only major force affecting the velocity of a falling object. Human behavior and the formation of archaeological sites are not like this. As early social scientists realized, there are many causes for the complex behavior and ideas of living humans (Vogt 2007). Similarly, as Binford (1976) was quick to note, there are many variables affecting the characteristics of archaeological assemblages, including human behavior in the past with all of its complexity, and the *milieu* of natural taphonomic forces acting on archaeological sites during and after there formation. Put bluntly, multivariate statistics are important for archaeologists because they help model a complex multivariate reality.

This section should not be taken as a philosophical attack on either the deductive approach to science or the utility of statistical hypothesis testing. I am both a card-carrying logical positivist and processualist archaeologist, and I firmly believe that the wide-spread use of probability statistics has radically improved the practices of archaeology. I am only trying to point out that multivariate pattern recognition statistics are useful for archaeologists because they offer ways of reducing data, and they are powerful ways of dealing with complex data that go beyond simple descriptive statistics. I am also trying to make the case that not all quantitative research in archaeology must result in the calculation of a p-value less than 0.05. Science is a fundamentally creative endeavor, and archaeologists should strive to be creative in their use of pattern recognition statistics.

The Boring Math Part: Multivariate Statistical Models and Assumptions

There are innumerable kinds of multivariate statistical approaches employed in the social sciences. This makes choosing a limited set to discuss in this chapter somewhat difficult. My choice is informed by two main research issues: (1) Conventional use of certain statistical techniques in archaeology and related fields and (2) the accessibility of statistical methods for researchers in terms of common computer software packages. Below, I discuss factor analysis (FA), principal components analysis (PCA), correspondence analysis (CA), and multiple regression / partial correlation, mainly because these techniques are available on every major statistical software package and because these have long histories of use in the field of archaeology. This discussion leaves out many potentially powerful approaches and other diverse members of the factor analysis and regression / correlations families of statistics. This section begins by discussing and comparing the statistical assumptions behind FA and PCA and closes with an examination of the statistical basis for multiple regression in archaeology.

Concepts and Statistical Assumptions Behind FA, PCA, and CA

FA and PCA are very closely related statistical techniques based on many common sets of mathematical foundations. As Baxter (2003) points out, often when the term FA is being used in archaeology, it is really referring to PCA. Likewise, Cowgill (1977) makes the case that there are no significant differences between the two. I am going to begin this discussion as if they were the same, using the term "factor analysis" to refer to both, and I will end by examining the more specific differences between the two.

When factor analysis came into common use in archaeology during the 1960s and 1970s, it was a controversial statistical approach, especially for statisticians with enough mathematical knowledge to understand the computation behind it. Factor analysis is a technique based in matrix algebra, which is used to analyze complex correlation matrices. Factor analysis is notorious for its computational intensity and it has only been widely available to researchers since the rise of powerful personal computers. No archaeologist can be reasonably expected to understand this matrix algebra well enough contribute to the math behind these statistical models. I would be lying if I said that I understood it completely myself. However, this is an example of a tool that may be utilized by archaeologists without total knowledge of its complex details. It is a bit like pointing out that archaeologists may use a laser theodolite without completely understanding the machinery involved in producing and measuring the properties of the laser. The effect of this situation places a burden on the archaeological researcher to understand the basic statistical assumptions underlying the approaches being used.

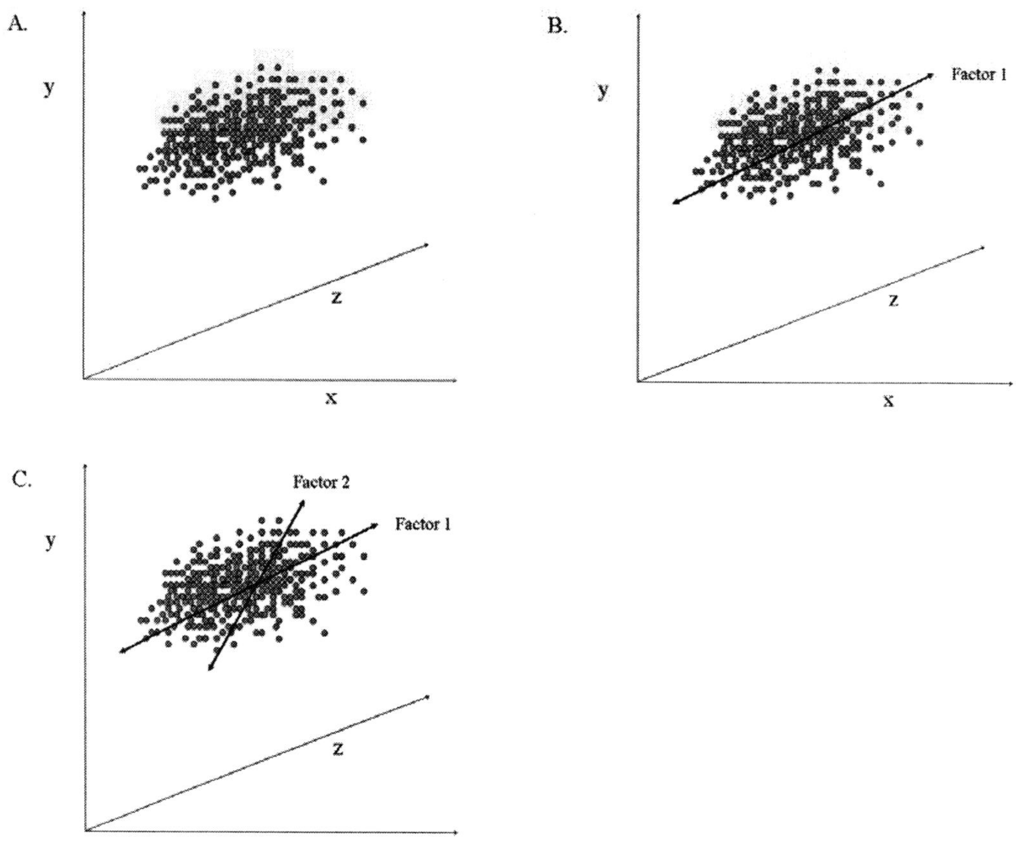

Figure 2. (A) shows the original swarm of data points in three-dimensional space. (B) shows the placement of the first factor regression line through the data, which best explains all of the data points. (C) shows the placement of the second factor regression line, which best explains the data not explained by the first factor. This processes continues with each successive factor addressing the data not explained well by previous factors.

In situations where many variables are operating, factor analysis presumes that, some of these variables will covary with one another. In other words, it works on the assumption that different variables will have relationships with one another, and that these relationships between variables are the result of some underlying dynamic of interest to the researcher. When different variables significantly covary, there must be some other variable or variables causing this relationship. The underlying dynamics that structure the covariation between variables are referred to as *latent variables*. They are latent in the sense that they are not visible in single observed variables, but are only seen through the covariation of several variables. Factor analysis hopes to discover latent variables by looking for patterns in the covariation of directly observed variables. This is the source of its power as a pattern recognition tool.

Factor analysis works by running a series of regression lines through data points in multidimensional space. These lines, referred to as factors, explain the data in best-fit fashion. The first line goes through all of the data points, and each subsequent line explains the data not explained very well by the previous line. Figure 2 shows this process with a fictional dataset. This example only shows three variable axes (x, y, and z), as it is hard to visualize more than three-dimensional multivariate space. Of course, factor analysis is not limited to three variables, but usually operates with many more than this. Each factor produces a series of regression scores for each variable that basically tells how strongly the variable covaries with the factor regression line. These scores are conventionally referred to as *eigenvalues*, and indicate how strongly a variable *loads* on a factor. Large positive eigenvalues indicate strong correlations of variables with a factor regression line, large negative values indicate strong correlations against a factor, and values near zero indicate weak relationships.

The examination of factor loadings is a way of recognizing the covariation between variables. When variables load strongly on the same factor, this indicates covariation, and points to a latent variable causing the relationship. This is especially the case when a set of variables all load strongly on a number of different factors. Frequently, these relationships are represented visually in two- or three-dimension scatterplots or hierarchical cluster analyses that show similarities in the loading of variables on major factors.

Factor analysis has two potential applications in analyzing data. Binford (1981) refers to these different applications of factor analysis as mode "R" and mode "Q." The "R" mode refers to the calculation of correlation coefficients for each of the variables. In other words, mode "R" is designed to look for latent variables through the covariation in observed independent variables. The "Q" mode refers to the use of factor analysis to understand how strongly the latent variables affect specific cases, and to use this information for the purposes of comparing and clustering cases. Mode "Q" involves the inversion of the correlation coefficient matrix to calculate scores for cases. This is frequently combined with other statistical approaches, such as hierarchical cluster analysis, in order to compare the characteristics of the cases. In dealing with the case studies presented later, both applications of factor analysis are used in each case, and I will use the "R" / "Q" distinction in talking about them.

Another common practice often performed in the course of a factor analysis is rotation. This is simply a process in which the factor solution rotated through the data points on which it was originally based in order to find the orientation that explains the data most parsimoniously. As with many things associated with factor analysis, this seems like an incredibly and confusing set of operations, and without the use of personal computers and statistical software, it certainly would be. Practically speaking, in most cases, performing a rotation is usually just a matter of clicking a box in the software program. The rotated solution is a more trustworthy reflection of the covariation between variables in the study, but usually the unrotated and rotated solutions do not differ very much. Many analyses, such as Binford's (1981), use the rotation as a way of generating regression scores for the cases in the study suitable for mode "Q" of the analysis.

As factor analysis is based on correlation matrices, it retains all of the regular assumptions underlying bivariate correlation and regression. These include continuous linear interval-scale data with more than two variables (and all of the other assumptions behind parametric statistics). This may sound like an imposing and confusing set of assumptions, but most large archaeological datasets meet them. For example, archaeological data representing both counts of objects in typological categories and standardized measurements of objects are both appropriate for factor analysis. The issue of whether data are linear and parametric is always complex, and should rightly be considered before any kind of statistical analysis. At the same time, with most archaeological datasets, it is probably safe to assume both of these unless there is some obvious *a priori* reason to think otherwise.

Let me briefly turn my attention to the unenviable task of distinguishing between FA and PCA in terms of their mathematical assumptions. For one thing, FA is the older of these two techniques, and to some extent, PCA represents an attempt to deal with some of the mathematical problems with FA. It is also the case that PCA is more widely used today, especially in fields like physical anthropology, where it has really become the industry standard. On the one hand, the two share most of their mathematical bases and generally produce highly similar results, especially with larger datasets. On the other hand, they do have one major distinction in terms of their mathematical assumptions that might sometimes have significant consequences in terms of the results of the analysis.

The major difference between FA and PCA is the orientation of the factor regression lines. One of the sources of controversy about FA has to do with possibility that the factor regression lines themselves might covary. This could be a major problem, because how can one use FA to recognize covariation between observed variables if the results of the FA also covary? The danger is that covariation in the factor regressions might make relatively minor relationships between variables seem much more significant than they really are, and *vice versa*. PCA addressed this problem by forcing its factor regression lines to be orthogonal to one another, thereby assuring that there is no mutual intercorrelation. In short, in PCA, the factor regression lines are assumed to be completely unrelated to one another, and therefore it is assumed that they must run at right angles, or be orthogonal. This is the nominal reason why PCA is generally the preferred variety of factor analysis today, especially in anthropology.

There is a sometimes-bitter debate over the importance or potential dangers of orthogonal factor regression lines within the statistical community (Vogt 2007). Advocates of FA point out that since factor regression lines are hopefully the result of interesting latent variables, there is no reason why they must be orthogonal, since there is no reason why latent variables themselves should be completely unrelated. As purely a consumer of this debate, I am compelled by aspects of both arguments. As Vogt (2007) points out, it is also usually the case that FA and PCA produce very similar results. For example, in a case study presented later in this chapter, I discuss a pattern recognition study of the Middle Stone Age lithic technology at Klasies River Mouth, South Africa, using PCA (McCall 2006). While I only reported the results of the PCA, I also did the same analysis using FA, and the two produced almost identical results. Vogt (2007) comments that this will usually be the case when large datasets are involved. Furthermore, using both statistical models to look at a dataset is a relatively easy option for most researchers, since it basically involves the clicking of a different box in statistical software program. Therefore, the debate over the superiority and advantages of FA and PCA usually has few major practical implications for anthropological research.

To summarize, FA and PCA will be most useful for archaeologists when: (1) Analyzing large datasets involving many variables and cases, and (2) these data are parametric and fulfill the assumptions of linear correlation and regression. These conditions are met by most archaeological datasets, which is why FA and PCA are powerful tools for pattern recognition in archaeology.

If there are clear reasons to think that a dataset does not meet the conditions of FA or PCA, there is another multivariate statistical method that is neither parametric nor linear called correspondence analysis (CA). The mathematics behind CA is similar to the other factor analysis techniques in relying on matrix algebra. The difference is that (1) CA was designed explicitly to deal with counts rather than including other kinds of interval data and (2) CA uses matrices of chi-square distances rather than correlation coefficients. The advantages of this include the potential to deal with non-linear non-parametric data, and in the past the calculations were dramatically simpler than the factor analysis techniques. The disadvantages of the technique are that it is much less powerful and lacks precision of the factor analysis techniques for the same reason that a non-parametric Spearman's *rho* measure of association is less precise than Pearson's r correlation. Though CA has been a very commonly used statistical technique in the past, there are relatively few situations today where it would be preferable to factor analysis techniques, given that most archaeological datasets are both linear and parametric, and given the ease of calculation using statistical software packages.

If there is uncertainty concerning the appropriate use of FA, PCA, or CA, one easy solution is to simply run all three using your statistical software. This is a quick way to determine if there are major differences in the results achieved using the different techniques. With most large datasets, there should be few major contradictions in the results of these three techniques. If there are large differences in the results, it may be a good idea to think harder about how the dataset under examination relates to the assumptions of these various statistical models. This is especially the case if the results of CA clearly contradict those of the FA or PCA. If this is the case, it may indicate that the dataset is either not appropriate for parametric statistics or the data are not linear.

Multiple Regression and Partial Correlation

Archaeological situations are frequently dominated by the dynamic interplay of multiple variables acting on one observed dependent variable. Multiple regression is a statistical technique for examining the overall effect of many different variables, and for recognizing which variables are having the strongest effect on one observed dependent variable. This stands in contrast with factor analysis, which is aimed at discovering latent variables causing covariation among the observed variables. Multiple regression does this by running a best-fit function through the scatter of observed data points in multivariate space. Simple linear regression does this with two variables, and the result is a line. Frequently, the linear regression equation is used to generate many useful statistics, including those related to correlation coefficients. Multiple regression works on the same principle, but includes many more dimensions. For example, the simplest multiple regression situation would involve three variables (two independent and one dependent) and the resulting regression equation would be a two-dimensional plane. Multiple regression obviously works with more than two independent variables; it can potentially have limitless number of independent variables. However, the results are impossible to display in graphical way.

By itself, the ability to put a regression equation through a scatter of data points in multivariate space is not that interesting, and this does not give much information of great use for archaeologists. The utility of this approach comes from the implications of the resulting regression equation for the relative effects of each independent variable on the observed dependent variable. For illustration, imagine the simplest multiple regression case with three variables. Each independent variable could easily be examined for its effect on one dependent variable using simple linear regression. The problem with this approach is that it lacks the potential to detect if one variable might be affecting the other, or which variable is having the strongest effect. Suppose that each individual regression shows a strong relationship between the independent and dependent variables. Multiple regression looks for the effect of both independent variables at the same time, and sees in multivariate space which variable has the strongest effect. Multiple regression equations can also be used to generate many of the same statistics as single linear regression, including multiple correlation coefficients and probability assessment.

A related technique is partial correlation, which involves examining the effect of specific variables while holding others constant. This is especially useful for finding the relative effects of two or more closely related variables on one independent variable. Mathematically, what partial correlation does is model two or more variables together using multiple regression and correlation, and then takes out the effect of certain variables on the others. This allows the researcher to statistically hold some variables constant while looking at the effects of others on the dependent variable. As Shennan (1997) points out, this is often practiced in reality in laboratory experiments, where certain variables can be controlled for through experimental design. Partial correlation offers a statistical way of doing this for archaeologists, who obviously cannot control variables operating in the past. Partial correlation offers another statistical technique for archaeologists dealing with variables that are often highly inter-correlated.

These statistical techniques are important tools for pattern recognition because they allow the researcher to look at many variables at the same time, and efficiently assess which variables are responsible for the most important observed effects. These techniques are most useful in situations where many closely related variables are involved. As with factor analysis, these techniques are parametric and assume that data are both linear and continuous. If data are not linear, it is still possible to do multiple regression, but the mathematics behind this become very complex, and this is often not allowed by most basic statistical software packages. A testament to this complexity is the work of Binford and Bertam (1977), who employed multiple curvilinear regression techniques working with animal bone density and anatomical part survivorship. The results were highly rewarding, but the mathematical complexity of even *reading* the paper is staggering.

Multiple regression and partial correlation are most frequently used in archaeology in the context of statistical hypothesis testing in basically the same way as simple bivariate regression. In fact, in collecting the case studies for this chapter, it was hard to find a good example of these statistical techniques used in a purely pattern recognition way. Indeed, the reader should be aware that if you must test a statistical hypothesis involving many variables, multiple regression and partial correlation are good ways of doing this. At the same time, they are generally under-utilized within archaeology in their roles as pattern recognition tools.

Learning by Example: On to the Case Studies!

At this point, the discussion of philosophy of science and mathematics is pushing the limits of most readers' attention. One of the best ways to learn how to use a new analytical tool is through the analogy of seeing someone else use it. For these reasons, I present a number of case studies using the statistical techniques discussed in this chapter on some types of material commonly dealt with by archaeologists. My hope is that these case studies will provide a framework for the reader in terms of (1) why the statistical approach was useful in a given situation and (2) how it was employed on a practical level by the researcher. These case studies should also illustrate some of my more opaque points in terms of the place of pattern recognition within archaeological science.

In this section, I present five important case studies covering the normal range of artifact types: (1) Binford's (1981) study of animal bones from the Early Stone Age (ESA) sites at Olduvai Gorge, (2) Bernardini's (2005) study of ceramic decoration in the later American Southwest, (3) Wilson's (1998) study of rock art motifs in Oceania, (4) my own use of PCA in studying the Middle Stone Age (MSA) lithics at Klasies River Mouth, South Africa, and (5) O'Connell's (1987) ethnoarchaeological study of the Alyawara using multiple regression.

I begin with Binford's (1981) case study, in part because of his pivotal role in bringing factor analysis techniques into the main stream of archaeological research. As Baxter (2003) points out, it was the success of Binford and Binford's (1966) paper that stimulated interest in factor analysis within archaeology. It is also the case that multivariate techniques for pattern recognition were a long-term interest of Spaulding (1977), and much of this work laid the foundation for Binford's early analyses. I choose to discuss Binford's (1981) faunal analysis because this is more like current applications of the factor analysis technique, and because it was among the first calculated using computers. It is important to credit Binford with his influence on the field both in terms of attitudes toward pattern recognition and the use of multivariate statistics.

Binford's Study of the Faunal Remains from Olduvai Gorge

This study comes at the close of Binford's (1981) highly influential book on the archaeology of animal bones. Most of the book is devoted to actualistic studies of modern animal bone assemblages produced by natural forces (e.g. modification by carnivores) and ethnoarchaeological studies of modern forager groups, such as the Nunamiut of Alaska. These are presented with the goal of illustrating the past dynamics responsible for the formation of archaeological bone assemblages, and constructing frames of reference for making inferences about past human behavior based on animal bone refuse. The book ends by using FA to look for patterns in the Olduvai ESA animal bone assemblages, and using the established analytical frameworks to make inferences concerning early hominid behavior.

Binford uses data from Leakey's (1971) summary monograph on the Olduvai excavations during the 1960s. Factor analysis was a useful analytical approach here because of Leakey's large dataset composed of many different variables and numerous cases. In performing his factor analysis, Binford treats the frequencies of various bone elements (presented in terms of animal part MNI, or a term later referred to as MAUs) as variables. Naturally, animals have many bone elements, creating a situation with many variables. Binford looked at 17 categories of anatomical parts for medium to large-size mammals. Olduvai is really a group of archaeological sites with many animal bone assemblages coming from different levels, which are treated as cases in this study. Binford examines the faunal data from 24 levels from 9 sites-- a relatively large number of cases. This is a situation commonly faced by archaeologists: Binford wanted to know about the covariation in the frequencies of various anatomical elements in order to look for the effects of latent variables, such as ravaging by carnivores or hominid bone-breaking for marrow, which are impossible to measure directly. In such a case, there are too many variables and cases to look for patterns effectively based on hunches or trying to eyeball every possibility with descriptive statistics.

Another interesting aspect of Binford's analysis is his integration of control samples into the factor analysis. In addition to the 24 archaeological cases, he also included 4 control samples-- 2 carnivore kill sites and 2 den sites. This is an important way of making sense of the results of the factor analysis, as it is known in these cases what the forces acting on the bone assemblages were, creating reference points for the other archaeological assemblages. Binford's analysis is composed to two basic phases: the "R" mode and the "Q" mode.

For mode "R" of the analysis, Binford recognizes 4 factors representing 4 latent variables affecting the frequencies of anatomical parts. The least dense anatomical elements that are most subject to destruction load strongly on the first factor. This leads Binford to recognize this factor as relating to bone destruction. The second factor groups animal parts most likely to be left behind at a kill site by carnivores, and the third factor groups parts most likely to be transported away from kill sites. The fourth factor groups parts which are most likely to covary as the result of large body size or old age of prey animals. Binford's interpretation of the meaning of these factor loadings is largely based on knowledge of the dynamic forces acting on animal bone assemblages outlined earlier in his book, and he also makes good use of his control samples in this part of the analysis.

For mode "Q" of the analysis, Binford uses these inferences of latent variables to examine the archaeological bone assemblages from Olduvai. In this phase of the analysis, Binford talks about a five-factor solution (though one wonders why the fifth factor was not discussed during mode "R" of the analysis). All of the cases score highly on the first factor, which represents destruction, indicating that all of the sites and the control samples show evidence for density-mediated attrition from numerous causes. Binford argues that the second factor represents transported assemblages, as the carnivore kill control samples do not score highly on this factor, but the den sites do. It is also interesting that relatively few of the archaeological assemblages score highly on this factor. None of the animal control samples score highly on the third factor, and Binford therefore takes this to represent activities of hominids. The Olduvai sites loading on the third factor have in common a pattern of high frequencies of scapulae, distal humeri, and distal tibiae, and low frequencies of distal metapodials. Binford sees this as the result of the removal of marrow-yielding elements from sites by hominids-- in other words, scavenging of animal resources left behind by the initial carnivores at kill or natural death sites. Binford sees factors four and five relating to the effects of carnivore kills (especially large kills in the case of the fourth factor), with the contribution of the elements missing from factor three sites, indicating the introduction of these parts by hominid scavengers. In summary, Binford uses FA in this study to examine the relative contribution of natural forces and hominid behavior to the characteristics of the faunal assemblages at Olduvai, and discovers an interesting archaeological pattern with important implications for early hominid behavior.

Binford's conclusions concerning early hominid scavenging have been quite controversial, though remarkably influential. In this study, whether or not one accepts Binford's explanation of the observed archaeological patterning, the utility of both his analytical approaches for quantifying animal bones and multivariate statistical methods for pattern recognition are clear. On a broader level, FA is a powerful tool for dealing with quantified faunal assemblage because of the obvious interesting in discovering pattern of covariation between categories of element. Most inferences about past behavior made from animal bone assemblages are at least partially based on element frequencies, especially when other variables (species of animal, age, sex, etc.) are included. As in Binford's study, by showing patterns of covariation in element frequency, FA is a method for identifying latent variables acting on animal bone assemblages that are impossible to examine directly, such as modification by non-human carnivores, density-mediated attrition, etc.

Binford's study also raises a few statistical questions that could easily be dealt with using modern statistical software packages, if it were done today. For example, Binford does not describe in great detail how much of the observed variation is explained by each factor. In FA, it is rare for factors beyond the second or third to explain enough of the variation to be very meaningful. In addition, as Baxter (2003) and Shennan (1997) both point out, aspects of Binford's statistical approach are more like PCA than FA, especially the rotation of the factor solution. These are minor issues, but there are currently better-defined conventions for dealing with them in terms of statistical nomenclature and what statistics are reported.

Bernardini's Work at Anderson Mesa and Homol'ovi

The next case study comes from Bernardini's (2005) study of the Hopi ceramic assemblages at Anderson Mesa and Homol'ovi-- groups of late prehistoric villages in northern Arizona. The research problem that Bernardini addresses concerns the dynamics of migration and aspects of social identity brought about by a period of drought-related population upheaval during the 14th century A.D. The specific archaeological question he examines is the movement of clans as indicated by the distribution of totemic clan signatures on the landscape and the transportation of ceramics as indicated by neutron activation analysis (NAA) of clay element compositions. Bernardini uses PCA to deal with the complex chemical signatures of ceramic vessels, composed of many different trace elements of interest in terms of sourcing.

Bernardini analyzed the amount of 14 different trace elements of 650 ceramic sherds from the Jeddito Yellow Ware (JYW) tradition from 15 sites at Anderson Mesa and Homol'ovi. This is clearly an appropriate case for the use of PCA, as it involves many different variables (amounts of the various trace elements) and many different cases (each individual sherd from the various sites). In this case, it is the covariation in trace element amounts that points to the latent variable of interest-- the source of the ceramics. Like Binford (1981), Bernardini also makes good use of reference samples by including chemical compositions of ceramics from 7 known locations, providing frames of reference for unknown samples. This approach allowed Bernardini to make inferences concerning the sources of ceramics at the Hopi villages in his study area.

Unlike Binford (1981), who was concerned with both modes "R" and "Q" in his FA, Bernardini is only really interested in mode "Q". The reason for this is obviously that mode "R" in this case would not provide information that was very interesting-- the patterns of covariation in element percentages for all sites considered together. In contrast, mode "Q" of the analysis tell how strongly the ceramics correlate with the various principal components, which provides a good basis for comparing and clustering ceramic assemblages on the basis of their source. Bernardini focuses on the first two principal components of the solution, which together explain 75% of the observed variation. Some of the reason for focusing on the first two principal components might be for ease of graphical display, as it is a lot easier to compose a two-dimensional rather than a three-dimensional graph. Even with only two principal components displayed, the graphs were quite striking in showing the association of ceramics of various villages in different locations. This information provided Bernardini with an important basis for making inferences about the flow of ceramics in his study area.

This case study also shows the value of multivariate statistics like PCA for dealing with complex trace element and sourcing analyses. Traditionally in archaeology, sourcing is done using other methods of pattern recognition. Often, the source of an object is inferred on the basis of the presence of one trace element found in a certain location but presumably not found at others. This method is sometime problematic because of the presumption that the signature element is not found at sources as of yet unknown to archaeological researchers. In other words, it is impossible to know what elements are present in the materials of unknown sources. It is also rare to find a case where there is one signature element that is unique to only one location. Another more complex approach is the comparison of amounts of many elements, which is the conceptual basis for Bernardini's study. This approach is problematic in that it is hard to recognize variation in the amounts of many different elements simultaneously that is linked to one specific location. Bernardini's study is innovative in using PCA to accomplish this task when trying to eyeball the covariation in element amounts with potential source locations in mind would be futile. Clearly, PCA provides a valuable statistical framework for handling the task of recognizing the complex patterns of covariation in element amounts present in different locations.

Bernardini's study is also partially based on a number of other statistical approaches for dealing with trace element amounts. For example, in establishing the initial reference groups, he uses a complex set of cluster analysis and probability assessment techniques. It also would have been better from a statistical standpoint if Bernardini had reported more information about the principal component solutions he arrived at, including a table of correlation coefficients and a list of eigenvalues. These were no doubt omitted in this paper for reasons of length and a larger focus on the social implications of ceramic sourcing information.

Wilson's Work on Petroglyphs from Oceania

I present the next case study not so much because it is a perfect model of how to use multivariate statistics to study rock art, but because it shows why this could be a very productive approach. As with many studies of Oceanian prehistory, Wilson's (1998) study is mainly concerned with examining questions of historical relationships between groups, migration, and interaction. In her study, Wilson uses CA to examine the relationship between rock art sites in various regions of the Pacific. The use of CA in this case may indicate Wilson's fears about whether her dataset meets the assumptions of parametric statistics and linearity-- the main reasons for using CA. This concern was probably not necessary in this case, as the frequencies of certain characteristics of rock art elements is an interval measure, and probably meets the other assumptions of FA and PCA models. Even given her conservative statistical approach, Wilson's use of multivariate statistics leads to many important insights about the prehistory of Oceania.

Wilson uses the frequencies of particular characteristics of rock art elements as the basis of her analysis. Her study focuses on the characteristics of anthropomorphs and geometric elements, and she treats the analysis of these two separately. In her study of anthropomorphs, Wilson examines 169 variables (characteristics of anthropomorphs, such as shape, presence of certain anatomical features, etc.) for 224 figures from 16 geographical regions. Obviously, 169 variables is a very large number, which is appropriate for using CA. However, these variables are not listed, and it presses the imagination a great deal to think of how there could be 169 easily identifiable features of each anthropomorphic figure. This study probably would have been better if it had reduced the number of variables to a more manageable set. For one thing, this sort of analysis does not have the ability to weight the importance of one variable versus another. I doubt that all 169 measured features of anthropomorphs are equally important or meaningful. Rather than coming up with a list of every conceivable minor aspect of variation in anthropomorph style, it might have been better to pick a smaller list of the most important stylistic difference. In general, one gets the impression that this number of variables was selected because of the *a priori* decision to use CA as a statistical approach.

Wilson uses two correspondence axes to examine the relationship between anthropomorphs in the various regions, and finds patterns of similarity and difference that generally confirm existing models of migration and settlement based on other ethnographic, linguistic, and archaeological data. Wilson's analysis of geometric shapes is based on the same design as with the anthropomorphs and came up with similar results. Wilson looked at 164 variables on 436 individual elements from the various regions of Oceania. Once again, a finer-grained focus on the most important aspects of variation in geometric shape style might have been preferable. I do not mean to take too critical of a tone with Wilson's work, despite some of the problems I pointed out, as her use of multivariate statistics to study rock art is quite unique and its seems to succeed in recognizing meaningful patterning concerning prehistory.

Let us stop for a moment and consider what the latent variables of interest are and how they relate to the covariation of observed variables. Wilson obviously hopes to identify latent variables concerning stylistic differences in rock art production based on differences of group identity stemming from patterns of migration and population history. In other words, she begins with an assumption that the many regional populations of the Pacific shared a common ancestor (probably in New Guinea) that had a relatively homogenous way of making rock art. Then, as the populations spread across the Pacific, diverse derived stylistic traditions of making rock art emerged, reflecting the cultural identity of group making it. In this scheme, the latent variable of interest is the cultural identity and historical context of the group making the rock art. This approach is squarely within the research paradigm of the anthropology of

Oceania, and Wilson does a good job of generating innovative quantitative approaches of using rock art to address historical questions.

It is not hard to see how a similar multivariate approach to rock art might be used to study different social dynamics in the past. For example, the kinds of rock art elements at a given site might vary for other interesting reasons. For example, different social or ritual contexts of rock art sites might lead to variation in patterns of rock art. In this case, a latent variable might be some ritual context in which a certain set of rock art elements was produced. Mode "R" could be used show the covariation in element frequencies and might point to related sets of imagery produced in common contexts. Mode "Q" could be used to identify different types of rock art sites with different contextual features. Multivariate statistics for pattern recognition important for rock art research because they offer the potential to see the covariation in frequencies of various rock art elements, which is one of the primary issues facing rock art researchers. While her research is not perfect, Wilson was definitely quite innovative in seeing the utility of multivariate statistics for dealing with covariation in element frequencies, and her study is fairly unique in this respect.

McCall's Study of the Lithics from Klasies River Mouth, South Africa

The next case study that I present is my own work on the lithics from Klasies River Mouth, South Africa (McCall 2006). This is an important archaeological site with remains including lithics, fauna, and early anatomically modern human skeletal material. Klasies River Mouth was initially important because it showed a very early presence of anatomically modern humans in South Africa. Later, debates developed concerning the nature of early human behavior on the basis of the lithic technology and faunal remains. One interesting aspect of the lithic assemblage is the emergence of the Upper Paleolithic-like Howiesons Poort industry, which is frequently related to the development of behavioral modernity in the Upper Pleistocene. The question I wanted to examine was whether there was continuity between the Howiesons Poort industry and later lithic technology present at Klasies River Mouth.

I performed a PCA using the categories of stone tools reported in the summary monograph of Singer and Wymer (1982) as variables and the culture-stratigraphic levels of the various caves at Klasies River Mouth as cases. In this study, there were 20 typological categories of stone tools and 8 culture-stratigraphic levels. The typological categories focused on strategies of core reduction and tool design, and mainly referred to differences between expedient, Levallois and blade core reduction strategies, and patterns of flake retouch. The culture-stratigraphic levels ranged included the Middle Stone Age (MSA) levels below to the Howiesons Poort, the Howiesons Poort, and the MSA levels above the Howiesons Poort.

Mode "R" of the analysis looked for relationships between the various categories of stone tools, and the results showed some clear patterns. The PCA found three main components: the first related to Levallois core reduction strategies commonly associated with the earlier MSA; the second related to blade core reduction strategies and microlith production typical of the Howiesons Poort; the third related to production of worked MSA points, and only this category of stone tools loaded strongly on the third principal component. This part of the analysis was useful in detecting latent variables affecting the covariation in stone tool frequencies related to basic approaches to lithic raw material and tool design.

In mode "Q" of the analysis, I used the varimax rotation to generate regression scores for the cases used in the study-- the culture-stratigraphic levels at Klasies River Mouth. These scores were useful for the purposes of clustering the cases and examining the question of continuity between the Howiesons Poort levels and those following them. The results of this part of the analysis showed that the MSA levels coming before and after the Howiesons Poort were much more similar to each other than any were to the Howiesons Poort. This is a provocative finding, suggesting that a radically novel strategy for stone tool production emerged at the end of the MSA, and then gave way to earlier strategies. This suggests that the emergence of the Howiesons Poort is not a clear signal of the origins of behavioral modernity, and it presents a case of the alternation of lithic industries similar to that seen in the Middle Paleolithic of Europe (e.g. Binford and Binford 1966).

PCA and FA are useful approaches for pattern recognition in lithic datasets because they have the potential to see covariation in both the frequencies of objects in various categories and between different kinds of measurement. The use of FA to study covariation in the frequencies of stone tool types has a very long history, extending back to early work of Binford and Binford (1966). Another interesting example of the use of PCA in studying stone tools is that of McPherron (2000), who uses the technique to examine the measurements of Lower Paleolithic hand axes. It is also the case that PCA and FA are flexible enough to deal with changes in the practices of lithic analysis and quantification and have remained useful for more than forty years.

O'Connell's Ethnoarchaeological Study of the Alyawara

The final case study I wish to present is O'Connell's (1987) ethnoarchaeological work with the Alyawara. This study used observations on living foragers in the Western Desert of Australia to understand the forces affecting site formation in the past. O'Connell was concerned with many characteristics of forager camp sites, including their location on the landscape, overall size, locations of various kinds of activity areas, and the locations of various kinds of refuse deposits—all things that are observable in the archaeological record. In order to understand to variables affecting these material characteristics, O'Connell made many observations on the human populations, demographic makeup of people living at a site, the kinds of activities that were done there, how long the site was occupied, and a number of other dynamics. This quantitative part of the study employs multiple regression because of interrelationship of many variables involved in shaping the various characteristics of camp sites, and the need to build models based on many variables.

I will briefly present one main example of the use of multiple regression in this study: modeling variables affecting the size of household activity areas. Here, the size of activity area refers to the size of the domestic space in which daily activities were carried out. These areas were cleared of vegetation and included important physical features, including domestic structures and hearths. This is the kind of information that archaeologists frequently deal with, and it is important to understand what human behavior is responsible for their various characteristics. In building a model to understand household activity area size, O'Connell needed to incorporate two variables to accurately understand the phenomenon: (1) the number of days for which the camp was occupied and (2) the total number of individuals occupying the household.

O'Connell's multivariate model of household activity area size has some interesting implications for archaeological research. First, he finds that activity area size relates directly with both household population size and duration of occupation. This is not an unexpected finding, simply showing that the more people that were present at a site and longer it was occupied, the larger the size of the resulting activity area. In this model, O'Connell makes the important connection between duration of occupation and household population size that could not be done effectively in separate, bivariate ways. It is also very important that this multiple regression model allowed O'Connell to hold each variable constant while examining the other. For example, this model shows that the effect of population size on activity area size decreases as population size increases. This basically means that amount of space per individual decreases as the total population increases. Similarly, holding population size constant, the effect of the length of occupation on activity area size decreases as the length of occupation increases. In other words, activity area size grows at a slower rate the longer a site is occupied. Both of these findings make a certain amount of intuitive sense, but it is important to be able to model them in a quantitative fashion, and knowing how these variables affect the physical layout of sites has very important implications for archaeological research.

O'Connell's (1987) use of multiple regression for ethnoarchaeological pattern recognition was innovative, and shows how the technique might be applied to other situations. For example, it is not hard to imagine the incorporation of other variables into a multivariate model of household activity area size or the expansion of the model to include other ethnoarchaeological cases. Indeed, O'Connell does make some interesting comparisons and contrasts at the close of his paper along these lines. More broadly, this ethnoarchaeological study is not unique in wanting to understand how many variables contribute to archaeological site formation. I stated at the outset of this chapter that many archaeological phenomena (especially the processes of site formation) are highly complex and multivariate in nature. O'Connell is also not alone in wanting the ability to hold the effects of certain variables constant while examining the effects of others. Controlling for the effects of specific variables is the key to sophisticated scientific archaeological model-building.

Summary Comments on Case Studies

It is my hope that these case studies have provided an *entrée* into the ways in which multivariate statistical techniques may be productively used for pattern recognition in archaeology. It is also my hope that seeing these statistical techniques in action might provide a more vivid demonstration of why (and perhaps how) these statistical procedure ccould be used. Along these lines, I highly recommend a more detailed reading of the original case studies on which this section has been based. Frequently, learning by example is easier than learning by abstract instruction, particularly for those of us whose analytical approaches are driven by the immediate needs of our research interests. I am also confident that a detailed reading of these case studies can provide a more compelling case for the use of certain statistical approaches than I have so far in this chapter.

Conclusion

In this chapter, I have tried to make the case for why certain multivariate statistical techniques could be extremely useful for pattern recognition in archaeological science. I wish to conclude with two main thoughts. The first is warning in terms of the relationship between scientific research problems and analytical methods. As Thomas (1978) warned almost thirty years ago, it is a mistake to put the "cart" of statistical methods and analytical procedures before the "horse" of specific research problems. In other words, the specifics involved in the investigation of the research problem should drive the selection of methods, not the other way around. This seems like a common sense statement, but is frequently violated as trendy theories, approaches, and methods come and go. This is no less true of statistical methods, which are often used for personality-driven reasons of convention rather than the requirements of research. More complicated statistical approaches do not equate with better science. My point in this chapter has been that there are often occasions where multivariate statistics might dramatically help the research process—not that they should be used in every situation all of the time. In this respect, Thomas points out the danger of "ritual" approaches to statistical analysis.

My second concluding thought is that too often statistical research is criticized, excluded, or ignored because of small mathematical problems. Statistical errors are inevitable, especially for archaeologists whose real job is understanding the past—not developing statistical models. As full-time consumers of statistical methods, mistakes are unavoidable, and we should acknowledge that these mistake range from very minor problems with no real implications for research conclusions to more serious and problematic errors. Most of the time, statistical mistakes in archaeology tend toward the former. Having made a great many mistakes in my own statistical analyses over the years, I know that these are not usually fatal flaws. The important point for this chapter is that researchers often fear the application of an unfamiliar approach because of the likelihood of mistakes. Statistical research in archaeology can be productive without being perfect, and fear of mistakes should not be an impediment to the development and application of innovative approaches.

Acknowdledgements

The author wishes to acknowledge George McCall, James Enloe, and Robert Franciscus for many helpful conversations and readings of this manuscript. Their assistance made this a better paper, but any mistakes are purely my own.

References

Baxter, M.J. (1994) *Explanatory Multivariate Statistics in Archaeology*. Edinburgh: Edinburgh University Press.

Baxter, M.J. (2003) *Statistics in Archaeology*. London: Arnold.

Bernardini, W. (2005) Reconsidering spatial and temporal aspects of prehistoric cultural identity: A case study from the Ameican Southwest. *American Antiquity,* **70**, 31-54.

Binford, L.R. (1968) Archaeological perspectives. In L.R. Binford and S.R. Binford (Eds.), *New Perspectives in Archaeology* (pp. 5-32). Chicago: Aldine.

Binford, L.R. (1976) Forty-seven trips: A case study in the character of some formation processes of the archaeological record. In E.S. Hall (Ed.), *Contributions to Anthropology: The Interior of Peoples of Northern Alaska* (pp. 299-351). Ottawa: National Museum of Man, *Mercury Series Paper* No. **49**.

Binford, L.R. (1981) *Bones: Ancient Men and Modern Myths*. New York: Academic Press.

Binford, L.R. & Bertram, J. (1977) Bone frequencies and attritional processes. In L.R. Binford (Ed.), *For Theory Building in Archaeology* (pp. 77-153). New York: Academic Press.

Binford, L.R. & Binford, S.R. (1966) A preliminary analysis of functional variability in the Near Eastern Mousterian of Levallois facies. *American Anthropologist*, **68**, 238-295.

Cowgill, G. (1977) The trouble with significance tests and what we can do about it. *American Antiquity*, **42**, 350-368.

Flannery, K.V. (1986) A visit to the Master. In K.V. Flannery (Ed.), *Guila Naquitz: Archaic Foraging and Early Agriculture in Oaxaca, Mexico* (pp. 511-520). Orlando: Academic Press.

Hempel, C.G. (1965) *Aspects of Scientific Explanation, and Other Essays in the Philosophy of Science*. New York: Free Press.

Leakey, M.D. (1971) *Olduvai Gorge: Excavations in Bed I and II*, 1960-1963. Cambridge: Cambridge University Press.

McCall, G.S. (2006) Multivariate perspectives on change and continuity in the Middle Stone Age lithics from Klasies River Mouth, South Africa. *Journal of Human Evolution*, **51**, 429-439.

McPherron, S. (2000) Hand axes as a measure of the mental capabilities of early hominids. *Journal of Archaeological Science*, **27**, 655-663.

O'Connell, J.F. (1987) Alyawara site structure and its archaeological implications. *American Antiquity*, **52**, 74-108.

Popper, K.R. (1959) *The Logic of Scientific Discovery*. London: Hitchison.

Shennan, S. (1997) *Quantifying Archaeology*. Iowa City: University of Iowa Press.

Singer, R. & Wymer, J.J.(1982) *The Middle Stone Age at Klasies River Mouth in South Africa*. Chicago: University of Chicago Press.

Spaulding, A.C. (1977) On growth and form in archaeology: Multivariate analysis. *Journal of Anthropological Research*, **33**, 1-15.

Thomas, D.H. (1978) The awful truth about statistics in archaeology. *American Antiquity*, **43**, 231-244.

Trigger, B. G. (1990). *A History of Archaeological Thought*. Cambridge: Cambridge University Press.

Vogt, W.P. (2007) *Quantitative Research for Professionals*. Boston: Pearson.

Wilson, M. (1998) Pacific rock-art and cultural genesis: A multivariate exploration. In C. Chippindale & P.S.C. Tacon (Eds.), *The Archaeology of Rock-Art* (pp.163-184). Cambridge: Cambridge University Press.

In: Pattern Recognition in Nanoscience...
Editor: Diane M. Perez, pp. 181-201

ISBN: 1-60021-715-X

Chapter 7

RULE-BASED RECOGNITION OF A POLYMORPHIC MESOAMERICAN SYMBOL

Yann Frauel[1,a], Octavio Quesada[2,b] and Ernesto Bribiesca[1,c]

[1] Departamento de Ciencias de la Computación, Instituto de Investigaciones en Matemáticas Aplicadas y en Sistemas,
Universidad Nacional Autónoma de México, México, DF 04510, Mexico

[2] Instituto de Investigaciones Filológicas, Universidad Nacional Autónoma de México, México, DF 04510, Mexico

Abstract

This chapter deals with the problem of detecting in archaeological drawings a symbol whose shape is extremely variable. We describe both a segmentation and a recognition technique designed to solve this problem. The extraction of the symbols is based on encoding the drawing as a set of discrete curves, and on properly grouping these curves. Because of the great shape variability of the symbol, we propose a recognition technique based on a set of rules. These rules characterize the global morphology of the symbol while allowing large variations between its actual instances. One of the rules is the existence of a single symmetry axis. We describe a comparison metric between curves in order to search for symmetries. The other rules used for the recognition verify the presence of an arch at the core of the symbol. We present recognition results obtained on various pre-hispanic images. The rule-based approach proposed and implemented here appears suitable to detect polymorphic signs, a common feature of Mesoamerican symbols.

1. Introduction

Mesoamerica is a cultural complex situated in Mexico and part of Central America, constituted by developed cultures that existed since around 1500 B.C., up to the sixteenth century. The Olmec, Teotihuacan, Maya, Zapotec, Toltec, Mixtec and Aztec cultures are

---

[a] E-mail address: yann@leibniz.iimas.unam.mx
[b] E-mail address: ernesto@leibniz.iimas.unam.mx
[c] E-mail address: oquesada@ifc.unam.mx

among the best known and studied. All of them show clear regional, cultural, and linguistic traits that single them out, nevertheless they conspicuously share a number of Mesoamerican traits. One of these is the extensive use of plastic shapes and media to represent divinity, an ancient tradition practiced everywhere in Mesoamerica.

In recent years, a series of comparative iconographic studies have focused on the images of god Tlaloc, and have identified in those and other related images, a number of main figurative motifs [1-4] and four signs [5,6] which show pan-Mesoamerican distribution. The four signs are highly polymorphic and show large variations in appearance and complexity, even though they seem to follow specific structural principles.

In this chapter, the polymorphic shape of one of those symbols is studied, defined and compared by pattern recognition techniques. Thus, this work deals with shape of symbols. Shape is an intrinsic property of objects. Whereas color, motion, and intensity are relatively simply quantified by a few well-understood parameters, shape is much more subtle [7]. Natural or painted shapes are incredibly complex. It is not clear what aspects of shapes are important for applications such as recognition. In order to facilitate the recognition of the shape of objects, it is necessary to select an appropriate shape representation. Most shape representations are based on two fundamental descriptions: contour and region oriented descriptors. Also, it is important to select a shape descriptor invariant under geometric transformations, such as: translation, rotation, and scaling. Several authors have used different techniques for shape analysis and recognition, such as sequential extraction of shape features [8], medial axis transforms, mirroring axes [9]. Pavlidis [10] reviewed these and other methods.

Ballard and Brown [7] proposed the template matching, which is an operation for finding out how well a template (a shape) matches a window of a given binary image. This binary image may be considered as another shape. Strachan et al [11] presented a method to compare the shapes of two fishes based on invariant moments, optimization of the mismatch, and shape descriptors. Another measure of shape similarity was presented in [12] by means of the shape numbers, which are related to the resolution of the digitalization scheme. These methods find the best matching between shapes.

Recently, other methods related to shape similarity have been proposed. Gunsel and Tekalp [13] propose a new shape-similarity metric in two eigenshape space for object/image retrieval from a visual database via query-by-example. Carlin [14] presents five different new performance measures for shape similarity retrieval. Mokhtarian and Abbasi [15] propose a method for shape-similarity retrieval under affine transforms based on the maxima of curvature scale space image. Mahmoudi et al [16] present a method for image retrieval based on shape similarity by edge orientation autocorrelogram. Torsello and Hancock [17] propose a skeletal measure of 2D shape similarity.

Due to the difficulty to detect Mesoamerican symbols, which are polymorphic, we propose a rule-based approach. Several authors have used this type of approach, such as: Ding and Young [18] have implemented a rule-based system to complete a shape from an imperfect contour. Another example of this type of systems is presented by Ahmed and Ward [19], who developed a novel rule-based system for thinning of characters. In this chapter, we define a few rules that describe morphological characteristics of the symbol. These rules are general enough to characterize the symbol in spite of its variability but they are specific enough to discriminate between correct and incorrect symbols in most of the cases [20]. In section 2, we present an overview of the problem we want to solve and the approach we developed. In section 3, we describe the preliminary stages of extracting shapes from an image and normalizing them. In sections 4 and 5, we explain in more details the implementation of the proposed rules, concerning the number of symmetry axes and the morphology of the shape respectively. In section 6, we give some recognition results. We present our conclusions in section 7.

2. Overview

2.1. Problem Statement

A considerable number of pictures and drawings of mesoamerican artifacts and monuments has been collected by archaeologists and iconographists who are interested in Mesoamerican cultures. These pictures can be organized in a database that can then be searched to extract images that have specific characteristics. In particular, in order to help the interpretation of the pictures, researchers are interested in finding all occurrences of a particular symbol. Until now, this search had to be done manually by an expert. For efficiency purposes, it is desirable to automatize the inspection of the database and the detection of symbols in each image. In this chapter, we focus on the study of the most basic of four newly identified transcultural mesoamerican signs [5,6]. We refer to it as symbol *One*. Figure 1 shows a few examples of the forms that this symbol can take. It is clear that there is a great variability of shape between the various occurrences of symbol *One*. Actually, we do not know in advance all the possible forms that the symbol can assume. Because of this polymorphism, it is impossible to define a single (or even a few) reference model of the symbol as required by a conventional template

Figure 1. Various examples of symbol *One* showing its variability of shape.

matching recognition process. It is thus necessary to characterize the symbol at a more conceptual level. The solution we propose in this chapter consists of defining a small set of rules that represent the global features of the symbol and that are verified by its various occurrences [20]. As can be seen in Fig. 1, symbol *One* can roughly be defined as a symmetric arch. Therefore the first rule we set is that a *One* has to possess one and only one symmetry axis. Second, we define several morphological rules that basically warrant that the central part of the symbol is an arch with suitable proportions.

It is important to note that our aim is to process a large number of images, each of which contains many symbols. It is thus essential that the detection algorithm we come up with be reasonably fast. For this reason, we use a boundary representation of our symbols rather than a region-based representation. Namely, the symbols will be represented by their contour curves. This approach is faster than a region-based approach because the recognition algorithm only has to process the boundary points of the symbol rather than all its inner points. Section 3, is devoted to describing how the symbols are extracted from an image and how they are represented using their boundaries.

2.2. Definitions

Throughout this chapter, we use a number of vocables to refer to various aspects of the symbols. We define here the precise meaning that we assign to each of these vocables. As a paradigm we consider that we are working on discrete (pixelized) digital images. A *point* M is a pixel of such an image. A *discrete curve* is a series of N ordered points (M_1, M_2,...M_N). Two consecutive points are not necessarily adjacent in the image; if they are not, they are assumed to be virtually linked by a straight segment. Since we only deal with discrete curves in this chapter, we also simply refer to them as *curves*. There exist two types of curves: a curve that has different starting and ending points ($M_1 \neq M_N$) is an *open curve*; a curve that has the same starting and ending point ($M_1 = M_N$) is a *closed curve* or a *loop*. Each closed curve can be seen as the boundary of – and therefore defines – a *symbol*. If the thickness of such a symbol is zero, then it is called *wire symbol*. Otherwise it is called *thick symbol*.

3. Preprocessing

3.1. Segmentation

The images we need to process are generally complex and contain a large number of strokes and interconnected elements, many of which are not relevant to us. The first step of the detection problem therefore consists in extracting symbols from the image so that they can be submitted to the rule verification process [20].

3.1.1. Assumptions on the Image

With an arbitrary image such as a photograph, the extraction of the symbols is in general a very difficult problem that largely exceeds the limits of this chapter. In order to make the problem tractable, we have to impose some restrictions on the type of images we use. The input image is therefore assumed to be a binary (black and white) line drawing. This means that it is constituted of black lines on a white background. In addition, we assume that the image is clean: lines are not erroneously interrupted and there are no extraneous connections due to noise. Such assumptions on the image are not unreasonable since it is a common practice among archaeologists to execute hand drawings of objects. Another possibility to obtain an input image in the desired form is to apply an edge detection algorithm on a more complex image and further filter the result to remove the noise. It is to note that the technique proposed in this chapter works as well on thick symbols as on wire symbols. However, for filled up thick symbols, it is necessary to perform an edge detection to extract the boundaries of the symbol before applying the algorithm.

3.1.2. Line Thinning

The first step for finding the symbols in an image is to extract the lines, since these will form the boundaries of the symbols. In an image, lines are typically several pixels thick. In order to follow them more easily, we first reduce their thickness to a single pixel thanks to a thinning algorithm. We therefore apply the Zhang-Suen algorithm [21,22] to the input image. This algorithm iteratively deletes boundary pixels while assuring that the topology is not modified (connections are not added nor deleted). After application of this algorithm, only the skeleton of the shape subsists. However this skeleton can still be two-pixel thick in some locations. In order to strictly reduce the thickness to one pixel, we apply our own final round of deletions: every pixel of the image is sequentially scanned and deleted if its 3x3 neighborhood corresponds to one of the eight cases given in Fig. 2.

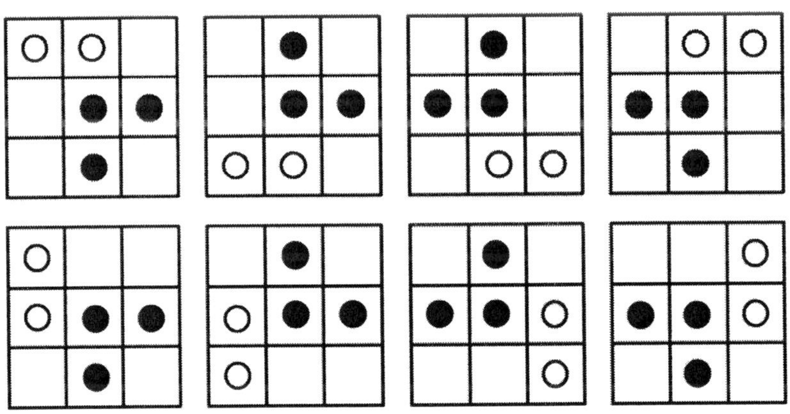

Figure 2. Configurations for the last step of the thinning algorithm. A dark circle symbolizes an object pixel, a white circle symbolizes a background pixel. Other pixels are indifferent. If one of these configurations is encountered, the central pixel is deleted.

3.1.3. Finding Nodes and Elementary Curves

In order to extract the symbols, we first need to inventory all the curves present in the image. To achieve this inventory, we first search the image for special points that are either the extremity of a curve or the intersection of several curves. We name such a point a *node*. In more precise terms, a node is a point that does not have exactly two neighbors in eight-connectivity. Once all the nodes are found, we look for all the curves starting at each node and we follow every curve until reaching another node. We term *branch* such a curve that connects two nodes. Note that, because each branch has two extremities, it is actually registered twice: once for each travel direction. After listing all the branches originating at every possible nodes, we erase them from the image. What remains is then the curves that do not have extremities nor intersections. These curves are therefore isolated loops and we register each of them as a branch. After this process, we know all the elementary curves contained in the image.

3.1.4. Finding Minimal Loops

The final step of the segmentation consists in extracting the symbols from the image. Of course there are many ways to group the elementary curves we extracted in order to form symbols. A simple yet efficient method is to extract minimal loops. These loops are obtained by using the right-hand rule: starting from the first branch, we look for every possible continuation (branches that start where the first branch ends) and we follow the rightmost one, that is, the one that forms the smallest oriented angle with the first branch. Now from this new branch again we look for the rightmost continuation and so on and so forth until the starting point is reached. The sequence of branches that were followed in the process are merged to form a minimal loop. They are then deleted from the list of branches in order to avoid duplicate use. We repeat this whole process from one of the remaining branches until there is no branch left. At this point, all the minimal loops have been found. These loops constitute the elemental shapes that we are going to test. It is important to note that, because every branch is registered in both travel directions, each curve actually appears in two different loops that are traveled in opposite directions. The result is that the process extracts both inside and outside symbols (see Fig. 3(a)). Besides, wire symbols are found as loops with thickness zero (Fig. 3(b)). We discard loops that are less than 15 pixels long since they are not significant.

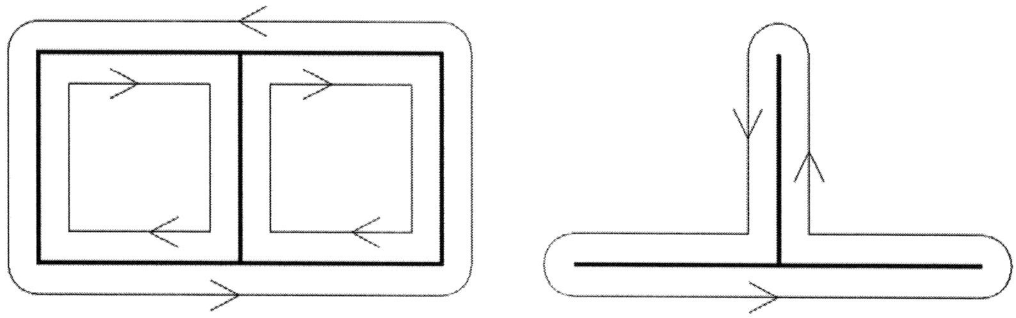

(a) (b)

Figure 3. Finding the minimal loops by following the rightmost connections. (a) Inside and outside loops are found. (b) Wire symbols are found as loops with thickness zero.

3.2. Normalization

In order to allow an easier comparison of the symbols and a standardization of the recognition rules, we need to normalize the position, size and orientation of the symbols. The normalization in position is simply done by centering the loop with respect to its centroid. Namely, the coordinates of the centroid are subtracted from the coordinates of every point of the loop.

For the normalization in size, we first determine the principal axes of the boundary loop. This is done by diagonalizing the covariance matrix of the point coordinates:

$$C = \frac{1}{N}\sum_{i=1}^{N}\begin{bmatrix} x_i^2 & x_i y_i \\ x_i y_i & y_i^2 \end{bmatrix}, \qquad (1)$$

where (x_i,y_i) are the coordinates of a point of the loop and N is the number of points. Note that the means of x_i and y_i are 0 because of the normalization in position. The eigenvectors of matrix C provide the directions of the principal axes and the square roots of its eigenvalues give an estimation of the extension of the symbol along these axes. We scale the symbol in such a way that its longest axis has a length of roughly 100 pixels. In other terms, if V_M is the largest eigenvalue of matrix C, we scale the symbol by a factor $100/\sqrt{V_M}$. The aspect ratio of the symbol is preserved while scaling.

Since discrete loops have a finite number of points, the scaling operation results in separating the points by a distance that depends on the original size of the symbol. In order to standardize the distance between points and also to reduce computation times, a subsampling of the loop is performed. That is, some points are discarded in a way that the distance between consecutive remaining points is close to 10 pixels. An additional benefit of this procedure is that it results in a lowpass filtering of the loop that eliminates local details that are irrelevant for the description of the global shape of the symbol.

Lastly, a normalization in orientation has to be performed. Since we are interested in symmetric symbols, it is natural to normalize the symbols with respect to their symmetry axis. Namely, we rotate the loop in a way that its best symmetry axis – if it has one – becomes vertical. The whole problem is to find the symmetry axes. Ideally, if a single symmetry axis exists, it should coincide with a principal axis. However, this coincidence does not always occur when the symbols are poorly drawn. Besides, a symbol can possess multiple and non-orthogonal symmetry axes. These additional axes therefore do not correspond to any principal axis, yet they should be detected as well. For these reasons, it is not possible to use the principal axes to find the symmetry axes. The following section describes the technique we use to detect axes of symmetry.

4. Symmetry Axes

For any particular axis, a specular version of a symbol can be obtained by mirroring the original symbol with respect to the axis. The axis is an axis of symmetry of the symbol if the specular symbol is superposable – without rotation – to the original symbol. In order to discover symmetry axes, we therefore need a way of measuring the similarity between two curves.

4.1. Comparison Metric for Open Curves

Let us first consider the comparison of two open curves A and B. This case is made easier by the fact that both curves have defined starting and ending points. Several comparison metrics have been proposed in the literature and one of the most popular is the Hausdorff distance [23]. Assuming that the two curves are optimally superimposed, the idea is to find for each point P on A the closest point Q on B. The Hausdorff distance between the two curves is the worst-case distance between P and Q, that is the largest of the minimized distances between a point of A and a point of B. A problem of this metric is that it is sometimes misleading as the distance between two very different curves can be low in particular cases. This problem happens because the succession order of the points is not taken into account. A proper comparison must consider that both curves are walked from start to end, without the possibility to go back. The so-called Fréchet distance [24] incorporates this restriction. It finds the largest of the minimized distances between the current point on A and the current point on B when performing the start-to-end walk. As an analogy, let us imagine that a man is walking on A and his dog on B. Then the Fréchet distance between the curves is the minimum size of the leash.

In this chapter we use a modified version of the Fréchet distance where we take into account the average distance of the walk rather than the largest distance [20]. We assume that the walks are optimized in order to minimize the average distance. The formal definition of the distance we use is as follows:

Definition 1 (modified Fréchet distance for discrete curves).

Let $A = (A_1, A_2 \dots A_M)$ and $B = (B_1, B_2 \dots B_N)$ be two discrete open curves with M and N points respectively. Let $\alpha : [1,K] \to [1,M]$ and $\beta : [1,K] \to [1,N]$ be two continuous and non-decreasing selections of the indices ($K \geq M$ and $K \geq N$). Then the *modified Fréchet distance* is defined as:

$$\delta(A,B) = \operatorname*{Inf}_{\alpha,\beta} \frac{1}{K} \sum_{i=1}^{K} \left\| A_{\alpha(i)} - B_{\beta(i)} \right\|, \tag{2}$$

where $\|\cdot\|$ stands for the Euclidean distance and K is the number of steps required to go from the starting point to the ending point on both curves.

In order to visualize the computation of the modified Fréchet distance, we can construct a bidimensional grid that represents all the possible associations between a point of A and a point of B. The columns of this grid correspond to the points of A while its rows correspond to the points of B. A node (i,j) represents a pair of associated points (A_i,B_j). A simultaneous walk on curves A and B is represented as a path from node to node, starting at node (1,1) and ending at node (M,N) (Fig. 4) [25,26]. Now each node of the grid can be weighted by the Euclidean distance between the corresponding points A_i and B_j. Finding the modified Fréchet distance between A and B is therefore equivalent to finding the optimal path between nodes

(1,1) and (*M,N*), i.e. the path with the minimum cumulative weight. Finding this path can be done efficiently by means of dynamic programming. The idea is that the progression on A and B is in only one direction. Going back is forbidden. So in order to reach node (*i,j*) of the grid, we have to come from one of the left or above nodes (*i-1,j-1*), (*i-1,j*) or (*i,j-1*). If we know the cumulative weights of the optimal paths that lead to each of these three nodes, we can select the node with the minimum cumulative weight. The optimal path to go to node (*i,j*) is the one that comes from this best neighbor. The cumulative weight of the optimal path that leads to node (*i,j*) is thus the sum of the weight of node (*i,j*) and the cumulative weight of its best neighbor. In this way, we can recursively compute the cumulative sum of every node, starting from low indexes nodes [20,27]. In order to compute the modified Fréchet distance between A and B, we need to compute the cumulative sum for every node of the graph until reaching node (*M,N*). This computation requires $O(MN)$ operations.

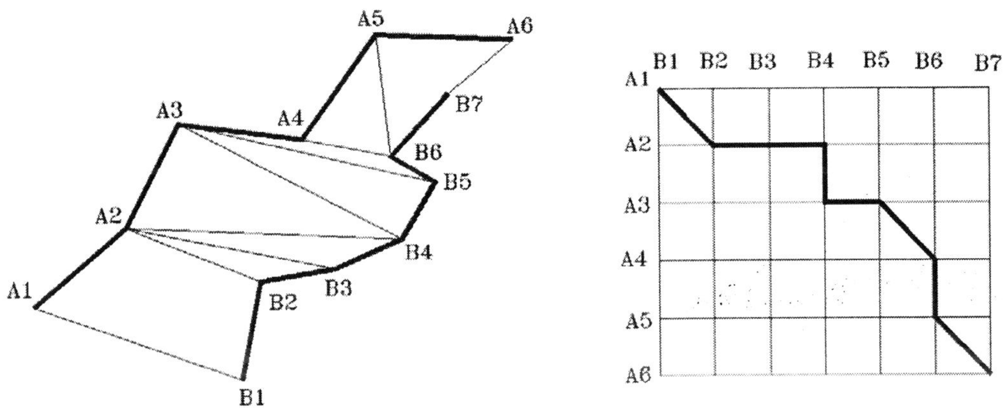

Figure 4. Grid representation of a simultaneous progression on two discrete curves.

Now the symbols we study are loops rather than open curves. This means that they have no definite starting or ending point. We therefore need to extend the comparison metric to loops.

4.2. Comparison Metric for Closed Curves

A straightforward idea is to split up a loop and consider it as an open curve by choosing an arbitrary starting point and walking the loop until going back to the chosen starting point. However each starting point provides a different "open" curve and it is not possible to know *a priori* which of them will optimize the comparison with another curve.

Now, if we want to compare loop A and loop B, we can choose – without any loss of generality – an arbitrary starting point on loop A and we obtain the corresponding "open" curve A'. On the other hand, we have to consider every point of loop B as a possible starting point. Depending on which point we choose, we can generate N "open" curves B' from B. We need to compute the modified Fréchet distance between each of these curves B' and curve A'. The resulting distance between loops A and B is the smallest of all the computed distances, that corresponds to the optimal starting point on B. Since there are N possible curves B', the total number of operations to compare two loops should be $O(MN^2)$. However, it is possible to optimize the comparison algorithm in order to reduce the total number of operations to $O(MN\log_2 N)$ [25]. Using this optimization, we are able to compute the distance between two 500 points loops in about 160 ms on a Pentium 4, 3.2 GHz computer.

4.3. Finding the Symmetry Axes

As mentioned above, a symbol that possesses an axial symmetry is identical to its specular image. However if the specular symbol is obtained by mirroring the original symbol with respect to an axis that is not the symmetry axis, then both symbols are not directly superposable. One of them has to be rotated in order to be perfectly superimposed with the other. The angle of rotation is the double of the angle between the mirroring axis and the actual symmetry axis. We use this fact to discover the symmetry axes of the symbol. The procedure consists of generating the specular image of a symbol with respect to an arbitrary axis – for example the vertical axis – and comparing rotated versions of this specular image with the original symbol. We will find as many correspondences as there exist symmetry axes.

In practice, the specular symbol is applied all the rotations between 0 and 2π radians with a step of 0.2 radians. For each rotation, it is compared with the original symbol using the modified Fréchet distance for loops as described in subsection 4.2. Every time the distance between the two curves is less than a threshold (heuristically set at 20 pixels), we add one to our count of symmetry axes.

With this procedure we finally obtain the total number of symmetry axes of the symbol. According to the rules defined in section 2.1, we discard the symbol if it has a number of symmetry axes different from one. Otherwise, we rotate it so that its symmetry axis becomes vertical, as mentioned in Section 3.2.

5. Arch Shape

Apart from the presence of a unique symmetry axis, the other prominent characteristic of a *One* is its arch shape. In this section we describe a few simple rules that are intended to verify this characteristic. In order to help this analysis, we first modify the coding of the symbols as described below.

5.1. Re-coding the Symbols

Since a symbol is usually not perfectly symmetric, we separately analyze the shapes of its right and left parts. We therefore need to split the loop that represents the symbol in two halves separated by the symmetry axis. The shape of each half will then be analyzed separately. In order to find were the loop should be split, we walk it to find where it crosses the symmetry axis. If more than two crossings are found, then we select the two of them that are farther apart from one another in terms of traveling the loop. The loop is then split in two open curves starting and ending at the crossing points. We refer to these two curves as *halves*.

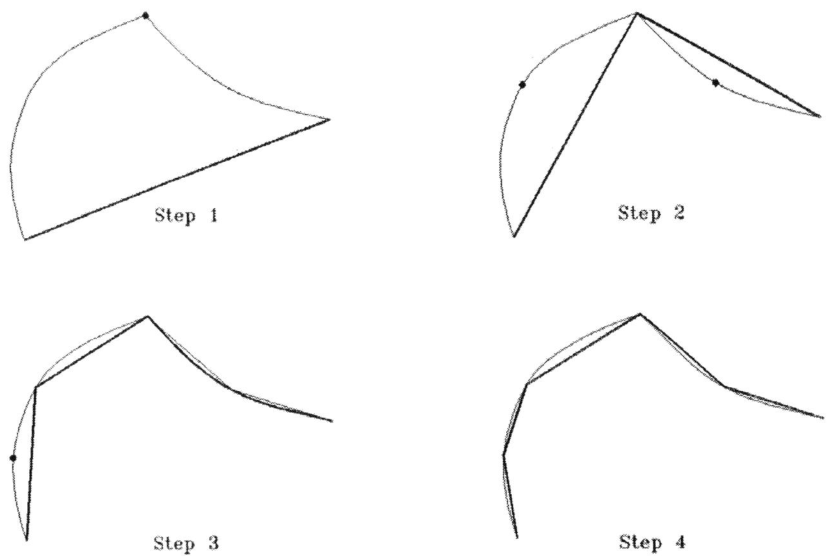

Figure 5. Regularization of a curve. The approximation starts with one segment and the farthest points are recursively added.

In order to eliminate small irregularities of the curves, we perform a regularization of each half. This regularization consists of approximating the curve by a small number of straight segments. This approximation is constructed recursively. We start with a single segment that directly joins the first and the last point of the curve, respectively M_1 and M_2. We then look for the point of the curve M_3 that is farthest away from the segment. The new approximation is thus the pair of segments M_1M_3 and M_3M_2. This two-segment description is a better approximation of the curve than the single-segment description. We repeat the process of splitting every segment in two until the distance between the segment and the farthest point of the curve is less than 10 pixels (Fig. 5). At the end, we are left with the regularized curve.

5.2. Extraction of the Components

As explained in section 2.1, a *One* is characterized by containing an arch in its central part, possibly prolongated by legs. This means that each half-symbol has to contain a semi-arch starting at the symmetry axis. This is only the case if its corresponding curve presents upper and lower bows near the symmetry axis (see Fig. 6). In order to avoid confusions, we use the term *arch* for the global shape of the symbol and *bow* for its boundary curve, although both words essentially refer to similar shapes. From now on, we consider the boundary curve that defines a half-symbol.

To perform the morphology analysis, we need to decompose the curve in its component parts, namely the upper bow, the lower bow and the leg. First of all,we obviously discard the curve if it is just a flat horizontal segment. Otherwise, when we walk either its upper or lower part starting from the symmetry axis, we eventually reach a non-horizontal section. The direction of inflection of the curve defines the direction of the bow. We keep following the curve while it continues in the same general direction (up or down). The traveled part of the upper (resp. lower) boundary, from start to end, defines the upper (resp. lower) bow. The non-traveled part of the boundary – if it exists – is called the *leg* (see Fig. 6). Note that this procedure can be performed in a similar way on wire curves.

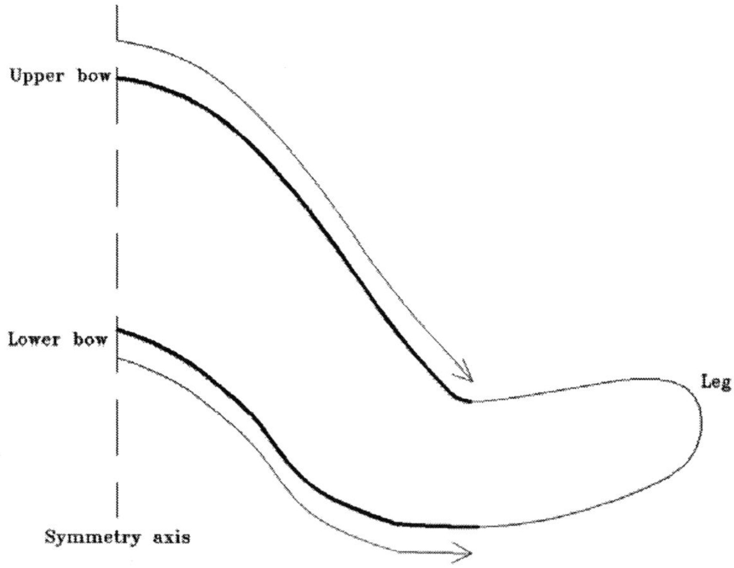

Figure 6. Extraction of the main components of the curve of a half-symbol.

5.3. Shape Analysis

Once the basic components of the half-symbol are extracted, we can define morphological rules that they have to satisfy in order to guarantee the arch property [20]:

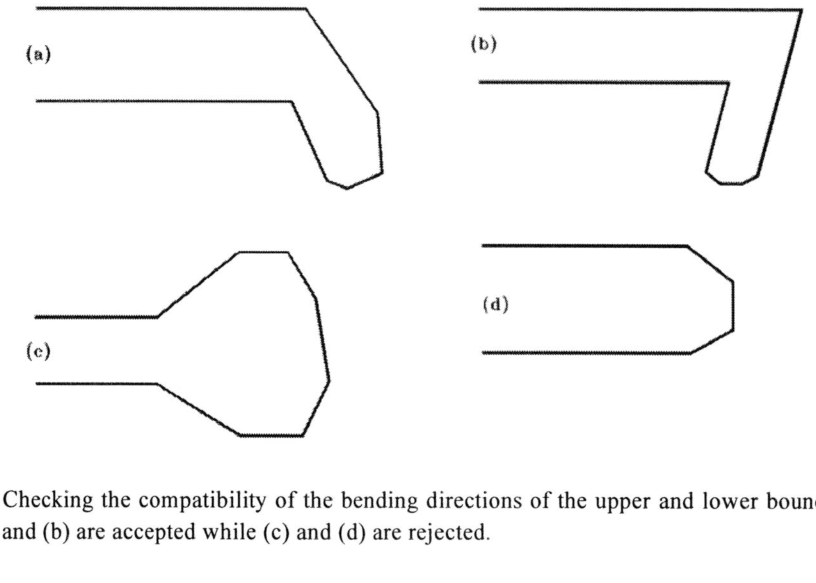

Figure 7. Checking the compatibility of the bending directions of the upper and lower boundary bows. Cases (a) and (b) are accepted while (c) and (d) are rejected.

1) **Identical bending directions:** the central part of the curve is a semi-arch only if the upper and lower boundary bows are bent in the same direction (up or down). If this condition is not fulfilled, the symbol is discarded (Fig. 7). Otherwise, we call outer bow (resp. inner bow) the bow that bends toward (resp. away from) the other bow.

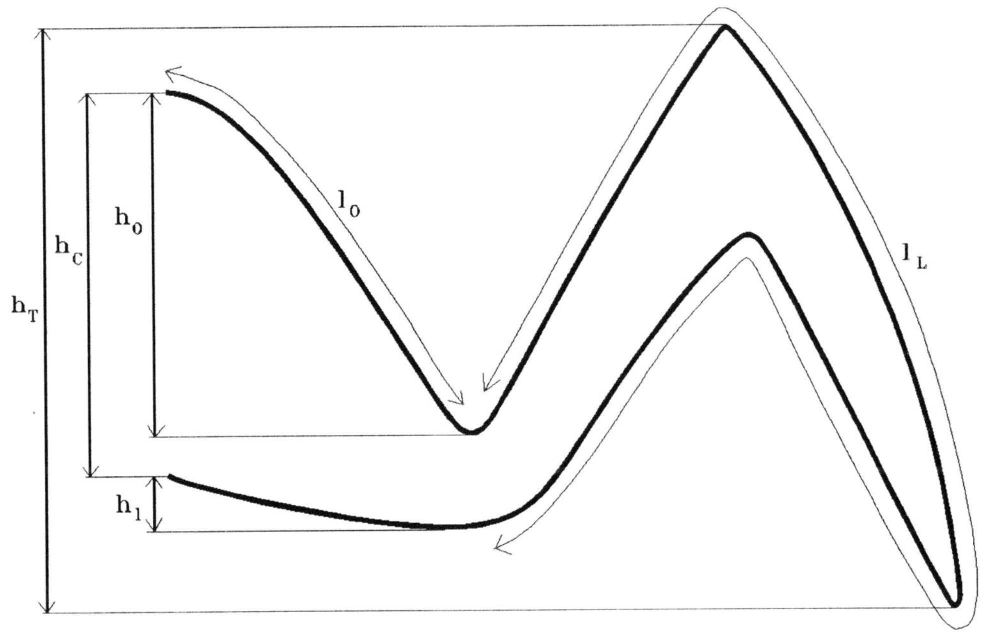

Figure 8. Definition of the parameters used in the morphological analysis.

2) **Height of the bows:** a curve is a bow only if it is sufficiently bent, that is, if its height is sufficient; otherwise it is almost flat. We define the height of a bow h_O or h_I as the difference between the height of its starting point (on the symmetry axis) and the height of its ending point (Fig. 8). The obtained bow height is to be compared to the total height of the half-symbol h_T (from its highest to its lowest point). The curve is rejected if one of the bow heights is less than 13 % of the total height, namely if $h_O < 0.13\ h_T$ or $h_I < 0.13\ h_T$.

3) **Compactness:** another characteristic of the *Ones* is their low compactness. Their length is usually significantly larger than their thickness. A good approximation of the thickness of a symbol is its thickness on the symmetry axis, that is its central height h_C (see Fig. 8). As for the length of the half-symbol, we approximate it as the length of the outer bow l_O plus half the length of the leg l_L (Fig. 8). The symbol is rejected if its length is not at least 60 % more than its thickness, namely if $l_O + l_L/2 < 1.6\ h_C$.

4) **Arch integrity:** the three rules above guarantee that the half-symbol we are studying has the shape of a semi-arch near the symmetry axis. As we already mentioned, the part of the curve that is not included in the outer or inner bows constitutes the leg of the symbol. This leg is supposed to be a continuation of the arch and, in particular, to lie out of the arch. In occasions, the leg can actually come back into the arch area and form an undesirable cavity that causes a degradation of the arch aspect. In this case, the symbol can no longer be considered as a *One*. In order to test this condition, we consider that the semi-arch is closed by the segment S that joins the ends of the outer and inner bows (Fig. 9). We then check if

any point of the leg lies inside the closed semi-arch. In practice, we tolerate small cavities that do no penetrate at distances greater than 25 pixels from the arch boundary. This tolerance has been found to be necessary in order to accept certain particular instances of *Ones*. However, any symbol with a cavity deeper than 25 pixels is rejected. Fig. 9 presents examples of accepted and rejected curves.

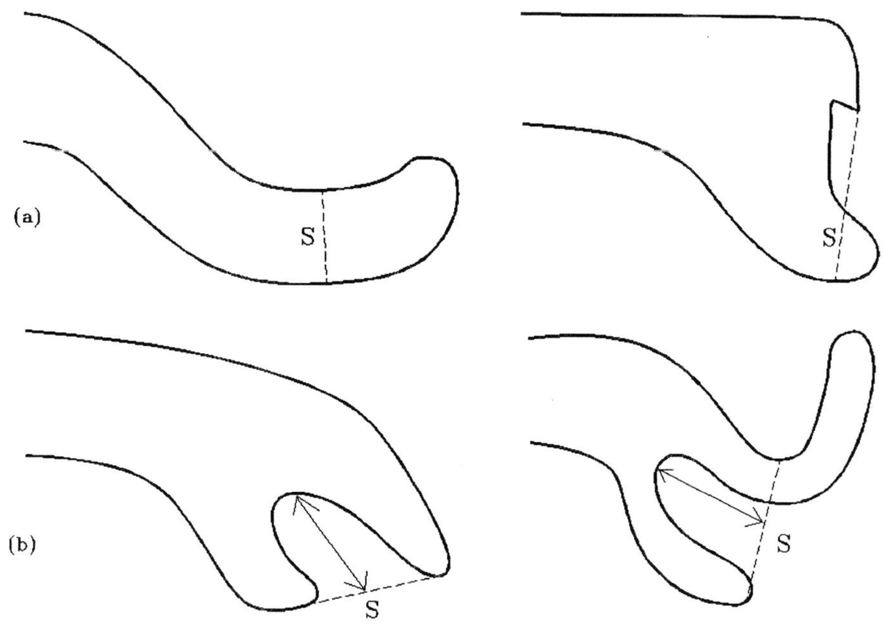

Figure 9. Testing the integrity of the semi-arch. *S* is the segment closing the arch. The arrows denote cavities deeper than 25 pixels. (a) Accepted curves. (b) Rejected curves.

When applying the four rules above, a symbol is discarded if any of its two halves doesn't comply. An additional rule is that both halves of a same symbol must be bent in the same direction (up or down). At the end, all the symbols that have not been rejected are considered as *Ones*. We have found that the application of these few rules provides a good recognition rate. In the following section, we present results obtained on various images.

6. Results

We now apply the proposed recognition technique to various images of increasing complexity. The first test is done on the image of Fig. 1 that contains several variations of *Ones*. All of them are correctly recognized. Now, Fig. 10 shows the recognition result of other examples of *Ones*. The symbols drawn in black are those recognized as *Ones*; the rejected parts are drawn in light gray. Only one of the symbols is not recognized, despite the great diversity of shapes.

Figure 10. Recognition of various instances of symbol *One*. The symbols recognized as *Ones* are drawn in black. Parts in light gray are rejected. One of the symbols is wrongly rejected.

Figure 11. Recognition of the *Ones* on a real composite drawing (Flat stamp from Mexico City). All the symbols are correctly detected.

Let us consider now actual cases. Figures 11 to 14 present real images of intermediate complexity. These images are due to Jorge Enciso [28]. Interestingly, Fig. 11 is almost entirely composed of *Ones*. All of them are correctly recognized. In Fig. 12, as in Fig. 11, most of the components are *Ones* and these *Ones* are well separated from each other. In that case, the algorithm also provides a 100% recognition rate. Figure 13 is slightly more difficult, because here several *Ones* are connected to other elements. However, all of them are still correctly extracted. Now Fig. 14 is the most complicated of the series, since it only contains a few *Ones* connected with several irrelevant components. Even in this case, we see that the *Ones* are recognized while the other elements are ignored.

Figure 12. Recognition of the *Ones* (continuation). Ceramic piece (malacate) from Mexico.

Figure 13. Recognition of the *Ones* (continuation). Cylindrical stamp from Veracruz.

Figure 14. Recognition of the *Ones* (continuation). Flat stamp from Oaxaca.

Figure 15. Recognition of the *Ones* in a more complex image. Huastec engraved stelae from Tepetzintla, Veracruz.

Figure 16. Recognition of the *Ones* (continuation). Zapotec engraved stelae from Monte Albán, Oaxaca.

Lastly, we apply the algorithm to more complex drawings. Figures 15 to 17 are given by Miguel Covarrubias in [1]. In Fig. 15, most of the *Ones* are recognized and only a couple of false alarms are triggered. In Fig. 16 as well, the *Ones* are satisfactorily extracted. Even if a few symbols are incorrectly classified as *Ones*, let us keep in mind that this is because we chose relaxed thresholds in order to accept *Ones* that are sometimes poorly drawn. This is a choice we made: we prefer to recognize most of the *Ones* at the cost of accepting some *non-Ones*, rather than wrongly reject many *Ones*. The last test is made with the highly complex image shown in Fig. 17. Again, because of our loose criteria, a few symbols are wrongly recognized as *Ones*. However, the most important is that practically all the *Ones* are recognized. This result is very satisfactory given the difficulty of the task.

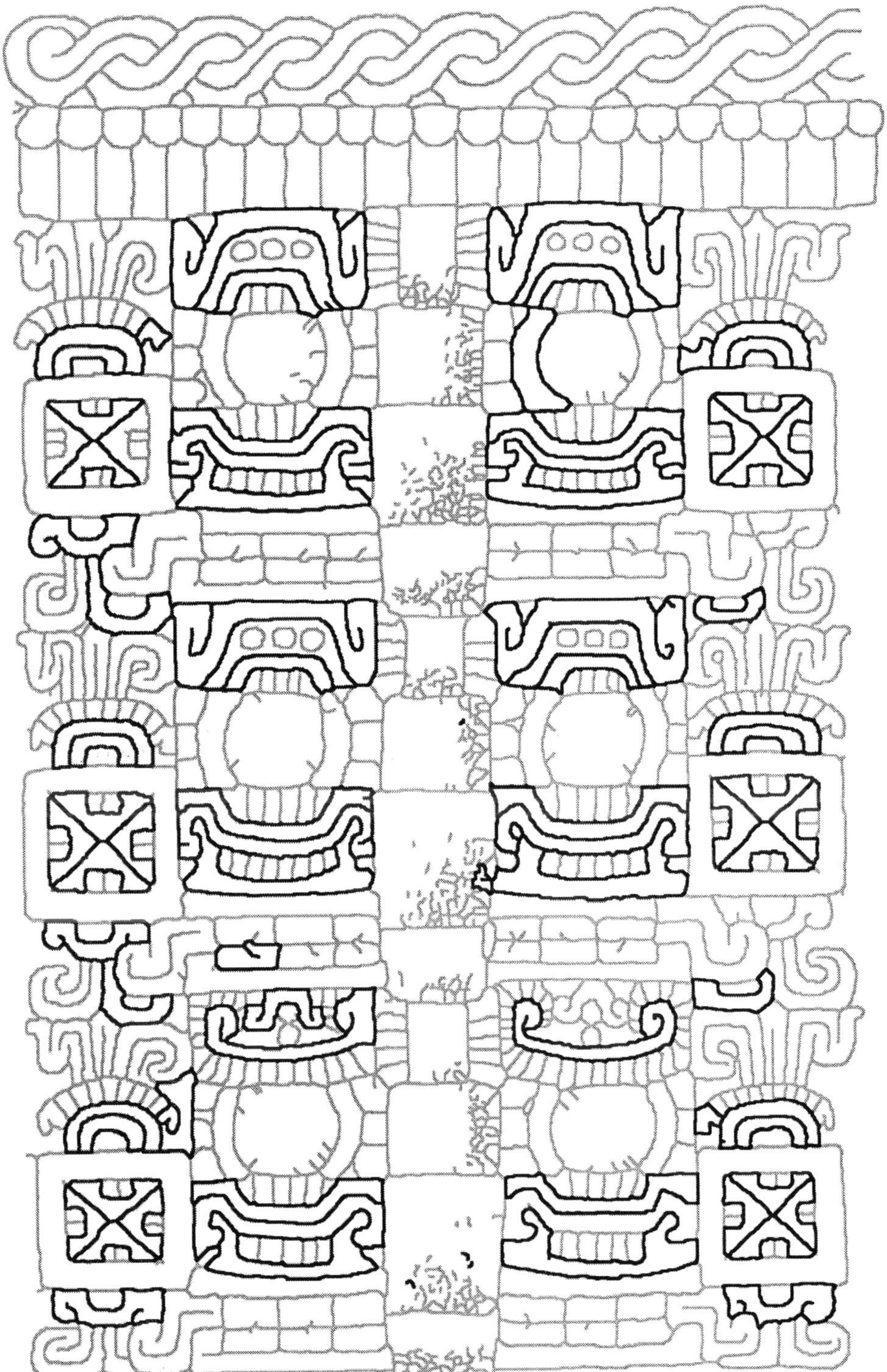

Figure 17. Recognition of the *Ones* (continuation). Maya masks of god Chaac from Uxmal, Yucatán.

7. Conclusion

In this chapter, we have presented a technique to recognize a Mesoamerican symbol that can assume a great variety of shapes. In order to deal with this variability, we have proposed an approach based on a series of rules that define the essence of the symbol and are common to all its occurrences. The rules we have used amount to verifying the existence of a unique symmetry axis and the presence of an arch at the core of the symbol. The number of rules is actually very small and simple enough so that the processing of an image is relatively fast (from a few seconds for a simple drawing up to a couple of minutes for very complex images). We have verified on actual archaeological drawings that the algorithm is able to deal with quite complex images. Actually, so far we have only used simple rules. The recognition process could probably be made more accurate by refining or adding new rules. In particular, an effort should be made to reduce the number of arbitrary parameters – such as thresholds – that are very sensitive and difficult to tune. Finally, it should be mentioned that the proposed algorithm is only able to recognize complete and unoccluded symbols. Dealing with interruptions or occlusions is a much more complex problem that remains to be solved.

Acknowledgments

The authors' gratitude is extended to Professor Rubén Bonifaz Nuño for his continuous support and encouragement in this work. In addition, the authors wish to thank Dr. Hermilo Sanchez Cruz and Lic. Yazmin Ocadiz Martinez for their useful scientific discussions. This study was supported in part by CONACYT, Mexico, Grant C-811-2000.

References

[1] Covarrubias, M. *Indian art of Mexico and Central America;* Alfred A. Knopf: New York, NY, 1957.

[2] Bonifaz Nuño, R. *Imagen de Tláloc: Hipótesis iconográfica y textual;* Universidad Nacional Autónoma de México: Mexico City, 1986.

[3] Bonifaz Nuño, R. *Hombres y serpientes. Iconografía Olmeca;* Universidad Nacional Autónoma de México: Mexico City, 1989.

[4] Bonifaz Nuño, R. *Olmecas: esencia y fundación. Hipótesis iconográfica y textual;* El Colegio Nacional: Mexico City, 1992.

[5] Bonifaz Nuño, R. *Cosmogonía antigua Mexicana. Hipótesis iconográfica y textual;* Universidad Nacional Autónoma de México: Mexico City, 1995.

[6] Quesada, O. *Tres Signos. Escritura antigua de Mesoamérica;* Universidad Nacional Autónoma de México: Mexico City, 2006.

[7] Ballard, D. H.; Brown, C. M. *Computer vision;* Prentice-Hall: Englewood Cliffs, NJ, 1982.

[8] Agrawala, A. K.; Kulkarni, A. V. *Comput. Graph. Image Process.* 1977, 6, 538-557.

[9] Wechsler, H. *Comput. Graph. Image Process.* 1979, 9, 246-266.

[10] Pavlidis, T. *IEEE Trans. Pattern Anal. Mach. Intell.* 1980, 2, 301-312.

[11] Strachan, N. J. C.; Nesvadba, P.; Allen, A. R. *Pattern Recognit.* 1990, 23, 539-544.

[12] Bribiesca, E.; Guzmán, A. *Pattern Recognit.* 1980, 12, 101-112.
[13] Gunsel, B.; Tekalp, A. M. *Pattern Recognit.* 1998, 31, 931-944.
[14] Carlin, M. *Comput. Vis. Image Underst.* 2001, 84, 44-61.
[15] Mokhtarian, F.; Abbasi, S. *Pattern Recognit.* 2002, 35, 31-41.
[16] Mahmoudi, F.; Shanbehzadeh, J.; Eftekhari-Moghadam, A.-M.; Soltanian-Zadeh, H. *Pattern Recognit.* 2003, 36, 1725-1736.
[17] Torsello, A.; Hancock, E. R. *Comput. Vis. Image Underst.* 2004, 95, 1-29.
[18] Ding, Y.; Young, T. Y. *Comput. Vis. Image Underst.* 1998, 70, 197-211.
[19] Ahmed, M.; Ward, R. *IEEE Trans. Pattern Anal. Mach. Intell.* 2002, 24, 1672-1678.
[20] Frauel, Y.; Quesada, O.; Bribiesca, E. *Pattern Recognit.* 2006, 39, 1380-1390.
[21] Zhang, Y. Y.; Suen, C. Y. *Commun. ACM* 1985, 27, 236-239.
[22] Parker, J. R. *Algorithms for image processing and computer vision;* John Wiley & Sons: New York, NY, 1994.
[23] Belogay, E.; Cabrelli, C.; Molter, U.; Shonkwiler, R. *Inf. Process. Lett.* 1997, 64, 17-22.
[24] Alt, H.; Godau, M. *Int. J. Comput. Geom. Appl.* 1995, 5, 75-91.
[25] Fuchs, H.; Kedem, Z. M.; Uselton, S. P. *Commun. ACM* 1977, 20, 693-702.
[26] Sederberg, T.; Greenwood, E. *Comput. Graph.* 1992, 26, 25-34.
[27] Eiter, T. and Mannila, H. *Technische Universitat Wien Technical Report* 1994, *CD-TR* 94/64.
[28] Enciso, J. *Design motifs of ancient Mexico;* Dover Publications: New York, NY, 1953.

INDEX

A

B

C

K

L

M

P

Q

R

S

T

U

V

W

Y